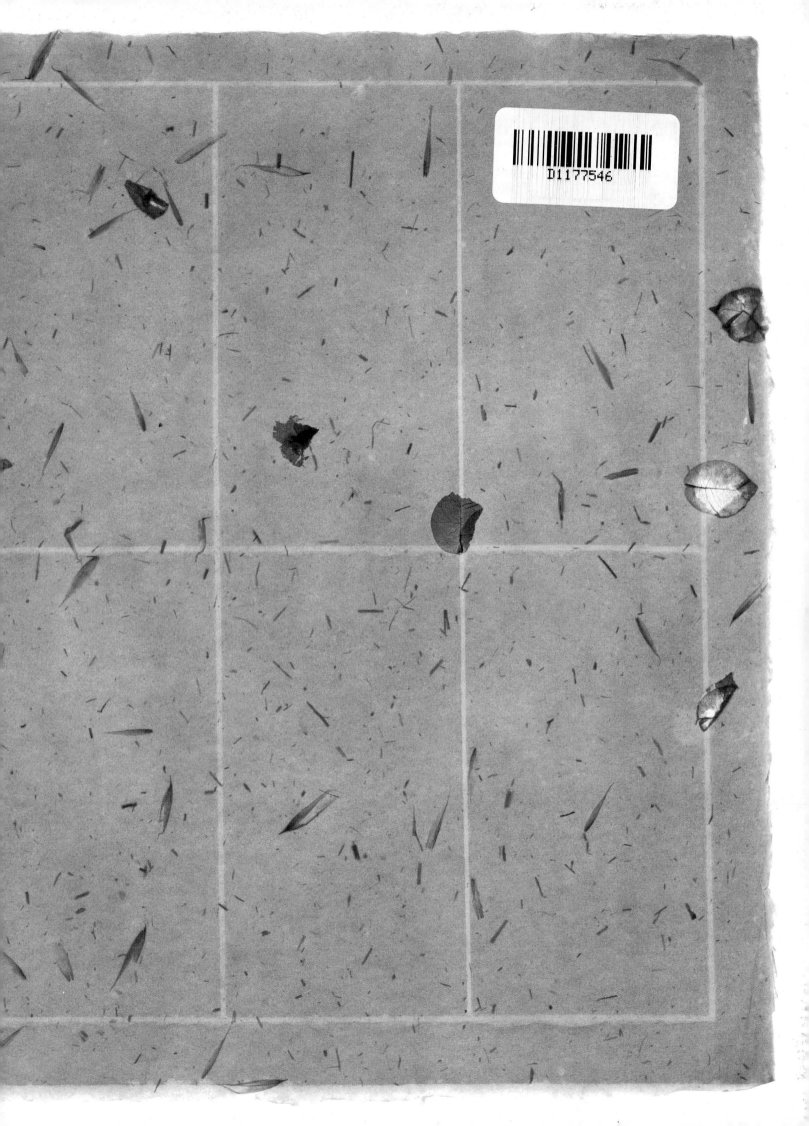

THE COMPLETE BOOK OF
PAPERMAKING

Josep Asunción

THE COMPLETE BOOK OF
PAPERMAKING

Josep Asunción

Library of Congress Cataloging-in-Publication Data

Asuncion, Josep.
  [Papel parramón ediciones. English]
  The complete book of papermaking / by Josep Asuncion.
    p. cm.
  Includes index.
  ISBN 1-57990-456-4
  1. Papermaking. 2. Paper, Handmade. I. Title.
  TS1105.A8513 2003
  676'.22--dc21
                        2002155637

10 9 8 7 6 5 4 3 2 1

Published by Lark Books,
a division of Sterling Publishing Co., Inc.
387 Park Avenue South,
New York, N.Y. 10016

Originally published under the title *El Papel*
Parramón Ediciones, S.A. Gran Via des les Corts
Catalanes, 322-324 08004 Barcelona, Spain
© 2001 Parramón Ediciones, S.A.—
World Rights

English Translation © 2003, Lark Books,
a division of Sterling Publishing Co., Inc.,
67 Broadway, Asheville, NC, 28801;
phone: 828-253-0467

Translation from the Spanish: *Eric A. Bye, M.A.*
Technical Consultant: *Claudia K. Lee*

Distributed in Canada by Sterling Publishing,
c/o Canadian Manda Group, One Atlantic Ave.,
Suite 105 Toronto, Ontario, Canada M6K 3E7

The Complete Book of Papermaking
Josep Asunción

Editorial Director:
Maria Fernanda Canal

Editorial and Image File Assistant:
Maria Carmen Ramos

Text and Coordination:
Josep Asunción
José Antonio García Hortal collaborated on
the chapter on fibers.

Exercises Carried Out By:
Josep Asunción, **Angels Arroyo**,
Teresa Collado, **Jordi Catafal**,
Lluis Morera, and **Oriol Mir**

Design:
Josep Guasch

Layout and Pagination:
Josep Guasch

Photography:
Nos & Soto

Illustrations:
Antoni Vidal

Production Director:
Rafael Marfil

First Edition, October 2001
ISBN: 84-342-2410-0
Legal deposit: NA-2161-2001
Printed in Spain

Con

HISTORY OF PAPER, 8

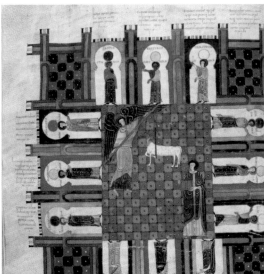

CHARACTERISTICS AND PROPERTIES OF PAPER, 18

ents

Introduction

THIS book explores a material that, despite its simple appearance, has been among the most versatile throughout history. Papers can be made that respond to all kinds of needs—from producing precious manuscripts to paper towels. It has also been used for such diverse purposes as industrial filters, delicate wrappings, jewels, scarves, clothing, works of art, wall partitions, and bags designed to hold heavy cement and plaster—not to mention the significant role it has played circulating from one hand to another in forms such as paper currency, checks, and stock shares.

Along with a long list of applications, there is a large inventory of papers bearing very different characteristics. In this book, we'll deal only with papers that have one thing in common: the fact that they can be made by hand. Accordingly, we'll focus on both the characteristics and the production of what is recognized as handmade paper.

Ever since I became involved in the world of handmade paper, I've met countless people who collect all kinds of things constructed from this material, including papers from distant countries, wrappings, notebooks, and stamps. These collectors almost always keep the papers intact and unused, out of respect for their natural beauty. My discovery of the profound respect and fascination that exists for handmade paper won me over and compelled me to learn more about this craft.

In this book, you'll get a close look at how to make paper by hand, including all of the technology involved: the components that go into paper, how the fibers behave, the use of materials and tools, and the process itself. Consequently, anyone interested in making paper will be able to do so with the aid of this book.

You can produce attractive paper using very limited resources, and if you want to continue experimenting, you can add to your resources bit by bit. Many of us in this field began with little equipment and added more as we grew in our interest and experience.

The book is divided into three sections. The first deals with the more theoretical aspects of paper, including a historical overview to give you a clearer understanding of why paper was invented and how it evolved over time. You'll also learn about the specific characteristics of a sheet of paper: its weight, surface, color, deckle edges, watermark, and so forth.

The second section of the book focuses on the technology of the craft: a description of materials and tools and how to use them, specific techniques for creating handmade paper (from a simple sheet to working with colored pulps or papers that incorporate natural materials), and, finally, a study of the fibers. I have based the final part of this section entirely on the fine work of Professor José Antonio García Hortal, who provided me with invaluable assistance (see the Bibliography on page 160 for information about his book).

Finally, the third section of this book contains a detailed, step-by-step study that involves a set of practical exercises to help you clearly understand the subject. Most of these exercises can be done in a workshop without elaborate setup or expense.

I believe this book will be useful to practitioners of the arts, teachers at art schools, curators, collectors, bibliophiles, and librarians. It will also be of interest to those of you who feel attracted to paper and want to interact with it more, as well as use it as a medium of expression.

Josep Asunción Pastor has a degree in Fine Arts from the University of Barcelona, Spain, with an emphasis in painting. Since 1983, he has combined his work as an artist with teaching. In 1987 he became Professor in the Arts and Crafts School of Barcelona, where he now teaches papermaking. He has also taught seminars, conferences, and numerous intensive courses in various locations, including the prestigious Rosa Sensat Summer School in Spain. He originally became interested in making paper because of its function as a ground for painting, and sought out courses with other artists in his region. As he progressed in papermaking, he broadened his contacts with specialists; notably Lluis Morera and Toni Capellades from the well-known paper mill, Ca l'Oliver; Gail Deery, Professor of Paper at Rutgers University in New Jersey; and the team of professionals at the Museu Moli Paperer de Capellades, headed by Victoria Raval and José A. García Hortal. Hortal is a Professor at the Polytechnic University of Cataluna, Spain, and is one of the most prominent specialists in the field.

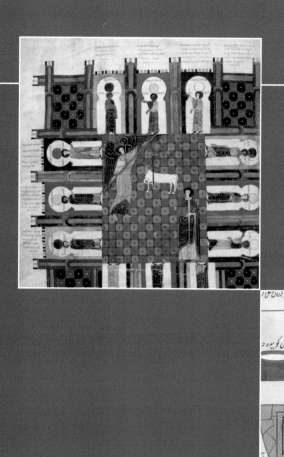

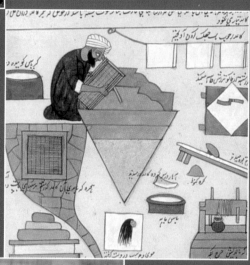

*T*HE invention of paper is credited to a civil servant named T'sai-Lun who worked under a Chinese emperor in A.D.105. By building on experiments that had been done with silk some 300 years before, he devised the simplest and best method for producing a sheet of paper. Thus, the first paper in history was born, which was an important advancement for humankind.

As the need for paper grew, so did technological advances, and the ways in which paper were made grew out of many sets of circumstances. In every place and time its composition, preparation, and treatment varied. One of the most important developments in this realm happened in the eighth century when the Arabs recycled cloth to make paper. Another major historical point was the invention of the printing press by Gutenburg in 1450, which ushered in an increased worldwide demand for paper. After this, the invention of the Hollander beater in the late seventeenth century, the multi-cylinder machine at the end of the eighteenth century, and the wood shredder in the mid-nineteenth century all contributed to large-scale industrial paper manufacturing.

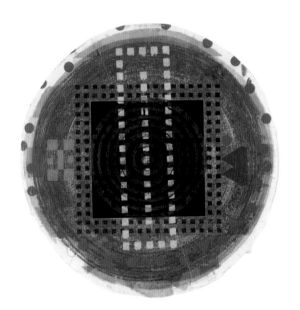

History
of paper

Antecedents of Paper

Humans have always felt the need to record their lives and express their longings in graphic form. Before they used paper, they did this on everything from the walls of caves to marble slabs, clay tablets, shell, bone, wood, and wax.

Historically, it's difficult to establish when one development ended and another began, since various methods have been used at different times. Among the antecedents of paper, the most common were tablets. Before the Asians used paper, tablets of wood, tortoise shell, and bamboo were incised with a hard awl. Since the marks made on these tablets couldn't be reversed, there was no way to correct the writing. Clay and wax tablets were more practical because their surfaces were more malleable, and they occupied less space. Tablets made of fresh clay left to dry were used in Mesopotamia. (In London's British Museum, for example, there are more than 20,000 clay tablets from the library at Nineveh constructed by Ashurbanipal.)

Wax tablets were used in classical Greek and Roman cultures, and they continued to be used until the Middle Ages. These tablets were composed of wooden or metal sheets covered with a layer of wax on which to write. The Romans referred to these as *tabula* or *tabella*. The wax tablets were written on with a pointed metal awl (*stilus*), the other end of which was either flat or spherical and could be used as an

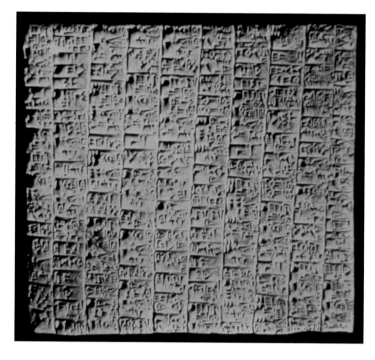

▲ Mesopotamian clay tablet from the royal palace at Ebla

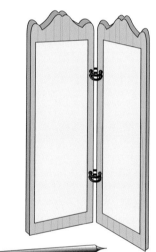

◀ Clay tablets with stylus

▼ Roman hospitality tablet from Badalona, Spain, ca. A.D. 98. Museum of Badalona.

eraser to make corrections. When a document included more than one *tabula*, it formed a *tabula dipticha*, *triptica*, or *polipticha*, depending on the number of tablets. (There was no wax on the first and the sixth tablets of a triptych, since they served as the covers.)

Metal plates made of bronze or lead were another surface for writing. Although complicated to write on, they proved to be extremely stable and have consequently survived since the days of ancient Rome, such as diplomas given to army graduates.

Stone and bronze slabs called *steles* are some of the largest and heaviest historical pieces bearing inscriptions. Conceptually, they are farthest from the lightness and portability of paper.

As civilization evolved, there was an obvious need for a lighter material that could be easily stored and transported. As a result, three fibrous materials with similar characteristics were developed independently in three locations around the globe: papyrus in the Mediterranean, pre-Columbian paper in America, and paper (as we know it today) in the Far East.

Papyrus is very similar to paper. The Egyptians used it as early as 3200 B.C. (In fact, the word paper is derived from the Greek word, *papuros*.) Pliny the Elder (A.D. 23 to 79) wrote about the different types of papyrus that were being made during his lifetime. Its use lasted until the tenth century, when paper production began to overtake papyrus.

The main documents of the Roman Empire were written on papyrus. For example, papal bulls written between A.D. 892 and 1017 still exist, giving us an idea of the importance of this material during the Middle Ages.

Papyrus is a plant that grows on the banks of rivers, particularly the Nile. Its Latin name is *Cyperus papyrus*. The properties of papyrus allow its leaves to be worked to form continuous, broad, smooth surfaces that are very similar to paper. Layers of leaves were placed parallel to one another at right angles before being beaten to release a natural liquid that bound them together in a single sheet.

Normally, sheets of papyrus measured about 12 to 20 inches long and 12 inches wide (30.5 to 50.8 cm x 30.5 cm). They were rolled up or kept flat in a volume. Because of the flexibility of papyrus, a long strip of joined pages could be rolled onto a wooden or bone rod to form a scroll. Some scrolls were as long as 20 to 40 yards (18 to 36 m) but the usual length was about 4½ yards (4.1 m). This presentation was maintained until the end of the fourth century when parchment became popular, and a bound codex format was used for writing.

▲ When a sheet of papyrus is backlit, it's easy to see the grid-like, reticular structure formed by the leaves of the plant.

To prepare the papyrus, its surface was polished using marble or agate. A quill called a *calamus* (reed), cut in the shape of a fountain pen, was used for writing on it. (Later, this tool was used on parchment, and quills made from bird feathers were used in Christendom.)

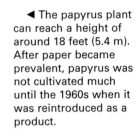

◄ The papyrus plant can reach a height of around 18 feet (5.4 m). After paper became prevalent, papyrus was not cultivated much until the 1960s when it was reintroduced as a product.

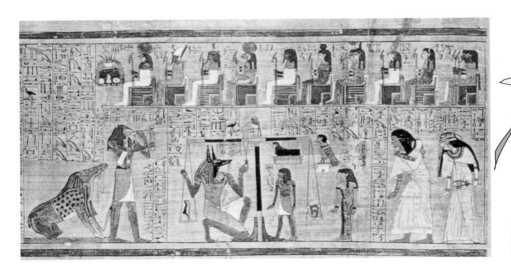

◄ Fragment of an *Egyptian Book of the Dead* written on papyrus. The most ancient papyrus documents discovered are over 5,000 years old.

Parchment

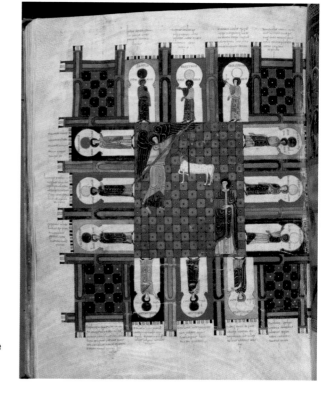

◀ Hartmann Schopper, *The Parchment Maker*, 1568; engraving.

▶ *Celestial Jerusalem.* Devotional on parchment from the National Library of Spain, ca. 1074.

The nomadic populations of Asia Minor were the first to use parchment, and we have documentary evidence of this fact from the ancient city of Pergamum from the years 258 to 197 B.C. Parchment quickly passed into widespread use; and, as a result, the library in Pergamum accumulated more than 200,000 volumes.

Persian and Greek kings used parchment for writing laws and answers from the oracles. The Roman philosopher and statesman Cicero (106 to 43 B.C.) claimed that the Greek poet Homer wrote *The Iliad* on parchment.

Parchment is produced primarily from the skin of goats, sheep, and calves, although it can also come from other animals. One variety of parchment was *vitela*, which yielded the finest and lightest type of parchment. It was prepared by removing the flesh and fat from animal skin using a scraper before soaking it, drying it, rubbing it with plaster, and smoothing it.

Until the thirteenth century, parchment sheets were usually made in monasteries. As making parchment became more widespread, guilds were organized. For a long time, parchment competed strongly with papyrus. The fact that parchment was more expensive than papyrus didn't hinder its growth at this time, because its superior durability and functionality overcame this obstacle.

The Latin name given to parchment over its long development period was *pergamēna*. As with papyrus, the writing implement was the *calamus*. Scrolls are the oldest parchment documents; and by the fifth century, both sides of the sheet were written on.

Since parchment was made from animal skins, it was difficult to produce the supply needed for libraries and other users. As a result, the parchment was recycled by washing and scraping it. Today, ultraviolet rays are used to read the content of the original codices that were obscured by cleaning the parchment and reusing it.

◀ Today, parchment can be purchased from businesses specializing in book binding. The whiter, cleaner, lighter, and more flexible the parchment is, the higher its quality.

Long before the Spanish conquistadors arrived in America, people experimented with writing on lightweight surfaces similar to those used by the Chinese and Egyptian cultures. It's believed that the Incas of the South American mountains used indigenous plant fibers or bark to make a paper-like material similar to papyrus. Remains of some of this material, dating from around 2100 B.C., were discovered in excavations carried out in Peru.

We don't know exactly when pre-Columbian paper came into use. The Central American Mayas discovered that bark fabrics used to make clothing could also be used for writing. These sheets were more durable than Egyptian papyrus because of their heavy texture.

This paper was made from a tree in the ficus family. The bark was removed in a single strip, left to soak for several days, and beaten on a flat trunk with another piece of wood until it became finely textured and elastic. This process can be used to make strips of soft, thin, malleable paper up to 18 feet (5.4 m) long by 28 inches (71 cm) wide.

When Hernando Cortes (the Spanish explorer and conquistador who conquered Aztec Mexico) arrived in Tenochtitlán (the ancient Aztec capital), pleated accordion books made from this paper existed. This paper was given to the Aztecs rulers as a tribute and used in their sacred rituals. A similar paper

1   2   3   4   5   6

▲ *Amate* paper made from red *xonote*. Its appearance is always a reminder that it comes from the bark of a tree, even though it's very soft to the touch.

The Otomi Indians live in close relationship to nature and the elements, and carry out magical and religious rituals using cut-out *amate* figures of mythological spirits.

Avocado god (1), the spirit of evil people (2), spirit of sick people (3), evil queen of the earth (4), apple god (5), spirit of god people (6). It's interesting to note that the evil spirits always wear shoes.

made from bark that is called *amate* is still made today by descendants of the Aztecs, the Otomi Indians of San Pablito, Mexico.

Amate paper, like its predecessor, is also made from the inner bark of a tree in the ficus family. In the *náhuatl* language, this tree is called *amacuahuitl* (tree) or *huun* (skin of the earth). These trees grow throughout Mexico and South America. The most common variety is the *xonote*. The best known, the red *xonote*, is the most widely known variety, and yields a brown paper. The mulberry *xonote* produces a higher grade of marbled paper with ochre veins. Other varieties of this tree produce more colors.

The Otomi Indians have been making amate the same way for over 500 years. It is used for religious and popular ceremonies and is exported in small amounts.

The *amate* paper of the Otomi is made like the first pre-Columbian papers, with a slight variation in the process. After removing the bark of the *xonote* tree and allowing it to dry in the sun for about two days, the strips are baked in limestone and ashes for several hours. When the fibers of the bark reach the point that they can be pulled apart by hand, the bark is removed to cool before being washed with plenty of water. At the end of the process, the fibers are wrung out in preparation for making a sheet of paper.

The fibers are arranged cross-wise on a smooth surface, like strands in a weaving. Then they are beaten with a rectangular piece of volcanic rock called a *moindo* or *mointo*, which is periodically moistened. As this is done, the fibers mesh to form a smooth, stable sheet that is placed in the sun to dry.

◄ A painted *amate* rich in color and symbolism

The Invention of Paper

Before the invention of paper as we know it today, a powerful Chinese general named Moung-Tian discovered Egyptian papyrus. Moung-Tian directed his artisans to look for a similar plant species in their area, but none of the plants produced the desired effect. Archeological discoveries near the great wall in Chinese Turkistan provide evidence of these experiments.

A technique that influenced the invention of paper was discovered by a Chinese man named Han Hsin (247 to 195 B.C.), who made a kind of felt from silk. He used the remains of coarse spun silk found in drums used for washing and bleaching the material, and placed it as filling between two layers of silk to make a kind of batting used to make shelters. Tablets discovered in the vicinity of the Gobi Desert from 100 B.C. contain this silk with texts written using a brush and lacquer-based paint.

In A.D.105, three centuries after the attempts ordered by Moung-Tian and the discovery of felt batting by Han Hsin, T'sai-Lun succeeded in creating the first paper in history by mixing plant fibers extracted from rags, fishing nets, mulberry-tree bark, nettle, and hemp.

To make these papers, the plant fibers were softened in lime water and left to ferment before they were crushed and ground to pulp using hand mortars. (In China at this time, it must have seemed natural to crush the plant fibers to produce pulp, since the Chinese were already crushing hemp to produce narcotics.) The resulting paste was mixed with water. Then a strainer made of bamboo fibers or cloth was submerged in the mixture to gather the amount of pulp needed to make a sheet of paper. Next, the strainer was hung on the walls of an oven to dry in the sun. Once the sheet dried, it was peeled off the strainer and burnished with a smooth stone. To waterproof the pages, they were coated with solutions made from an alga or plant juices.

Due to these discoveries, T'sai-Lun set up the first paper factory in history in Mongolian Turkistan.

▼ Chou Ling, *Portrait of Tsai-Lun*, 1964. Marius Peraudeau collection.

In the Orient, where many plant species were available for making paper, paper production from rags and recycling fibers from footwear and other materials was quickly abandoned. These papers were characteristically white and delicate.

◄ T'ien-kung k'aiwu, *Handmade Paper in China*, 1634.

The four main phases of papermaking can be observed in this series of four illustrations that accompanied a dissertation by Sung Ying-hsing: preparing the fiber, forming the sheet, pressing, and drying.

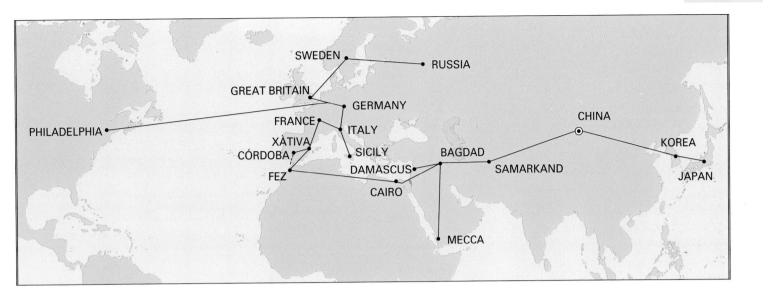

The discovery of paper increased the accumulation and dissemination of information. Consequently, the need for paper increased quickly.

The spread of paper began in the Chinese province of Hunan where it was invented in A.D. 105, and it traveled to other settlements in China until it reached Turfun in the north around A.D. 400. Then it began a journey along the Silk Road in the caravans of Persian and Assyrian merchants that reached as far as Central Asia.

Papermaking reached the Arab Muslims through another series of events. In A.D. 751, the Muslims prevailed over the Chinese in a battle near Samarkand (in present-day Uzbekistan). Among the Chinese prisoners taken were papermakers who revealed the secrets of their trade to the Muslims in exchange for privileged treatment.

The Arabs were a very learned culture, and their artists and intellectuals quickly took advantage of this invention. Over a period of time, Samarkand became a major production center, assisted by the local cultivation of flax and hemp. The paper industry that had been monopolized by the Chinese, moved westward via the Arabs, who set up paper mills in all of Asia Minor and North Africa.

In A.D. 793, a tremendous paper factory was constructed in Baghdad, Iraq. From Iraq, papermaking spread to Mecca and Cairo. (A Persian traveling through Egypt in A.D. 1035 recorded that many goods found in bazaars were wrapped in paper, indicating how widely it was used.) Papermaking eventually reached Fez in Morocco, and by A.D. 1200, Fez had 400 refining mills in operation.

Several technical innovations were introduced by the Arabs. First, they used recycled rags to make paper (a technique that the Chinese initially tried before using plants); second, they made metallic mesh strainers for handling the pulp; and third, they used pastes made from the starch of wheat flour as glues. The Arab papermaking industry underwent so much development that, by the tenth century, it produced more paper than papyrus. The Arabs also made ancient-looking papers by dyeing them yellow and brown with saffron and sycamore sap.

▲ Paper traveled on the Silk Road through Asia until it reached the West. From Samarkand, the spread of paper reached Europe through to Islamic culture. From Europe it was exported to America. Paper was known throughout the world by the seventeenth century.

▼ Manuscript of Cachemira illustrating the traditional trades and businesses from *The Paper Maker's Trade*, ca. 1860.

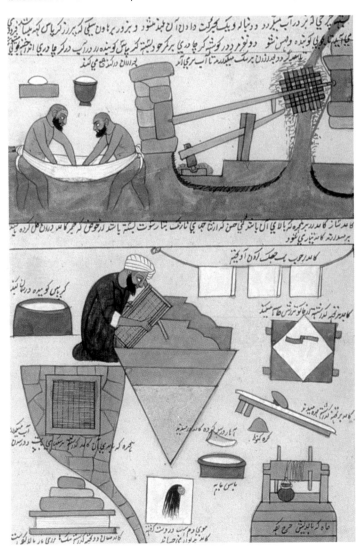

As Arabs moved through Europe via the Iberian Peninsula, they spread their knowledge of papermaking. Although there is no proof, it is believed that Cordoba, Spain (a part of Spain that was invaded by the Muslim Moors in the eighth century), was the first city in Europe to undertake papermaking. Some historians claim that the library of Caliph Al-Hakam II (part of the palace of Cordoba) contained more than 400,000 volumes.

Next to Cordoba, Toledo was the most important city during the Moorish domination of Spain, and it had paper mills known as "rag mills." The library of Ripoll in Toledo contained papers that originated in Cordoba. At the same time, there is evidence that there were paper mills in the Spanish cities of Xativa and Cataluña.

The first clear historical information about the details of a paper mill is from A.D. 1065 concerning a mill on the outskirts of Xativa. The mill employed 20 workers. This town became famous for the quality of its mill papers made from flax grown in Valencia.

▲ A sample of paper from Thailand. In many Buddhist countries, paper is used as a ritual element. Prayers are written on papers that serve as offerings to Buddha, and geometrical figures with symbolic meaning are depicted in gold or bright colors.

Papermaking was introduced in France around A.D. 1189, and by A.D. 1230, it was found in Italy—first in Genoa, and later in Bologna and Fabriano. Papermaking was later introduced to the island of Sicily. With assistance from the Italians, the first German paper mill was built in Nuremberg around A.D. 1400. About a century later, paper was introduced in Great Britain, and toward the end of the sixteenth century, paper had reached Russia, North America, and Sweden.

The use of paper spread slowly through Europe because the price was as high as parchment for a long time. Also, the diffusion process was slowed down by nobles of certain European countries who held prejudices against Jews and Arabs who manufactured paper.

The introduction of paper to Asian countries proceeded more quickly than in Europe. Papermaking in Japan began as early as A.D. 610 through contact with Korea. A Japanese cleric introduced paper into his country, using mulberry bark as its base.

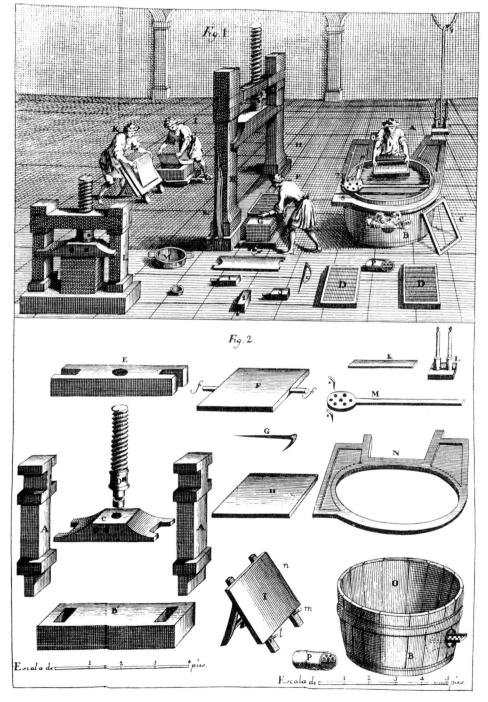

◀ Page 11 of the treatise of M. de la Lande, *The Art of Making Paper* (*Arte de hacer el papel*), 1768

Until the end of the eighteenth century, the only paper that existed was made by hand in traditional paper mills. There were a great number of mills concentrated in certain places. For example, in the small river basin of Cataluña, Spain, 80 mills were in operation. But, with the invention of machines that could mass-produce paper and the resulting industrialization of the mill, traditional mills closed down one after another until they were practically nonexistent. Today, machine-made paper dominates the entire Western market.

Nevertheless, in countries where this industrial explosion has been the greatest, there is considerable interest in reviving this craft—not simply for the sake of its human and cultural legacy, but also for its creative potential. Today, papermaking has taken a completely new form: many artists are working individually or in small teams to open up new roads by bringing back indigenous papers, experimenting with new materials, creating finer products in response to demands, and spreading the craft through teaching.

There has been a surge in interest among all sorts of artisans, especially in highly developed countries—perhaps as a reaction against industrial products and the overwhelming bombardment of large-scale commercialism that is part of

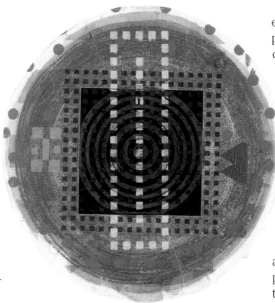

▲ Alan Shields, *Shields' shield*, 1972. Handmade relief and paper. Private collection. In this work, color has been directly applied to the paper pulp while it was being made.

everybody's daily existence. In the present worldwide culture, which is developing largely through virtual reality such as movies and the Internet, people long to relate to materials in a more natural way without the mediation of machines. The pleasure of touching paper that retains the mark of the hand that made it—expressed through uneven edges, texture, and irregular beauty—is a humanizing influence in the center of hectic contemporary life.

At the same time, there have always been parts of the world where paper has been made throughout history: Japan, the Philippines, Thailand, Ecuador, China, and Nepal. Paper made in these countries now travels through commercial networks to reach the world's major cities, where businesses specializing in handmade paper have become popular.

Many artists are now using handmade paper as a medium, not merely as a ground for their works. Paper, like any other material, has so much potential for expression that it is fascinating in and of itself.

◄ Joyce McDaniel, *Letter to a Young Sculptor*, 1996. Paper and metal. Private collection. Paper is used in contemporary sculpture to communicate its own messages.

▼ The influence of handmade paper on commercial paper products is increasingly evident. This kind of paper has become popular because of the consumer's desire for nature-oriented products.

◄ Akira Kurosaki, *Captured Time 84-1*, 1984. Kozo paper and bamboo, sumi ink, and cotton threads. The works of Kurosaki combine sculpture and painting. They have a meditative appearance produced by the subtle lightness of the Oriental paper.

D ESPITE the best efforts of industrial papermaking, no machine-made paper has been made that conveys the look and feel of handmade paper with its unique tactile and visual qualities.

Every paper, whether handmade or industrial, is defined by the following characteristics:

- a name;

- a composition: 100% cotton, 50% kraft + 50% eucalyptus, etc;

- a use: drawing, watercolor, decoration, etc;

- a weight (90 grams, 120 grams, 360 grams, etc.), which makes it paper, file stock, or cardboard, and gives it a certain thickness;

- a size that is defined according to standardized names (Couronne, Coquille, etc.);

- a color: natural, red, tobacco, violet, etc.;

- a surface finish: vellum, laid, engraving, satin, glossy, irregular, textured, etc.;

- deckle edges: long, stringy, irregular, rustic, etc.;

- a watermark;

- special characteristics: direction of the fibers, special glues, porosity, load, added materials, etc.;

- presentation: loose pages, packets, pads, by weight, etc.;

- other characteristics: opacity, hardness, strength, flexibility, water resistance, etc.

Characteristics and properties *of paper*

What is Paper?

What we call paper is actually a thin sheet produced from the physical bonding of previously hydrated fibrous materials, mostly cellulose.

When speaking of paper, it is always referred to in sheets. Any other presentation is given other descriptive names, such as paper pulp, paper maché, and so forth. A sheet of paper can have any desired dimension and thickness—from small, thin cigarette paper to a long, continuos roll—but it is still a single sheet without a break.

A sheet of paper is always formed in the same way, regardless of its type. Paper is physically bonded without agglutinating agents. (Sizing is added only for water resistance.) The fibers become tightly intertwined by suspending them in water and dripping the suspension onto a flat surface. Once the water has been drained from the sheet, it is usually pressed before it is dried.

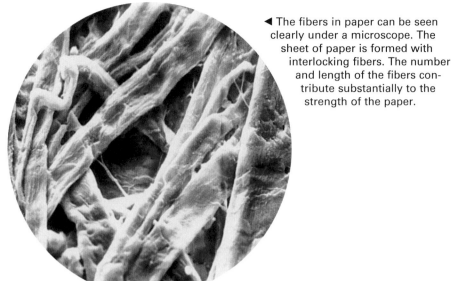

◀ The fibers in paper can be seen clearly under a microscope. The sheet of paper is formed with interlocking fibers. The number and length of the fibers contribute substantially to the strength of the paper.

▲ If you tear a sheet of paper and look very closely at the frayed edge at the tear, you can see the fibers of the paper without the aid of a microscope. Do this with different types of paper (shopping bags, newspaper, toilet paper, etc.), and compare the size of the fibers.

◀ A wasp makes its own kind of paper by constructing its nest using pulp that it makes from plant fibers and saliva. The physical and chemical processes that this pulp undergoes, as well as its final appearance, make the wasp a true paper artisan!

The way in which a sheet of paper is formed is the same as making felt. Like felt, paper doesn't have a woven structure, but is a product of matting the fibers together. When minuscule cellulose fibers are soaked with water, they become flexible, allowing them to interlace. Because the fibers are porous, they act like a sponge and remain in suspension, resulting in an even distribution of tightly interlocked fibers.

When the fibers dry, they lose their flexibility, they contract, and they harden. That's why a sheet of paper, which is so fragile when it's wet because of the hydrophilic quality of the cellulose, becomes extremely strong once the water is removed. For instance, think of a paper cement bag that can hold up to 110 pounds (50 kg) without tearing, while withstanding blows and tension. When that same paper was saturated with water as it was being made, it was as soft as bathroom tissue submerged in water.

Even if the sheet of paper contains substances other than the fibers, it is still considered to be paper because it has a cellulose structure. Those other substances can be sizing, coatings, pigments, and even added foreign materials such as flowers or threads.

What is Handmade Paper?

Paper is considered to be handmade when each sheet is created according to traditional methods employed by the papermaker.

The traditional method of making paper consists of submerging a mould (a frame covered with a screen) and a deckle (a second frame which is placed on the mould to contain the pulp) inside a vat that contains the fibers in an aqueous suspension. The mould is filled with this suspension (the pulp) and covered with the deckle before it is removed from the vat to drain off the water. The draining process is facilitated by gently rocking the mould and deckle back and forth. Once the water has drained off, the deckle is removed from the mould, and the wet sheet is placed onto a coarse piece of cloth or felt. The sheet is covered with another cloth and is pressed to finish squeezing out the remaining water. The process of placing the sheet between two cloths is called *couching*. This procedure is repeated as many times as desired to form a uniform pile of sheets.

Once the water has been removed, each sheet is removed one by one from its couching cloth or felt. In this state of dryness, the sheets are already much stronger than before. Next, the sheets are allowed to dry naturally in the sun, in the air, or on drying rack. They can also be dried artificially using an iron, fans, heaters, or other means. If necessary, the sheets can be pressed again to create a smoother surface.

Each fiber that makes up paper is composed of a tube with inner ducts and pores on the surface, like a plant sponge. Not all fibers are the same; they vary in width, regularity, porosity, surface texture, shape, and so on. Nevertheless, what they all have in common is their hydrophilic nature, or their ability to absorb water and remain in suspension without sinking to the bottom or floating to the top.

The Direction of the Fibers

Cellulose fibers and water used in paper-making integrate perfectly when the fiber adapts to the movement of the water. When the water is at rest, the fibers arrange themselves in a suspended pattern; and when the water moves, the fibers move with it in a longitudinal direction. The final direction that the fibers take must be kept in mind when creating paper for certain purposes—such as paper made for printing a book—since the direction of the fibers effect the paper's strength and how it takes ink. When paper is produced industrially or mechanically, the machine picks up or deposits the pulp onto the strainer in the same direction. As a result, the paper's fibers are aligned.

In contrast, the traditional method of making paper by hand disperses the fibers in many unpredictable directions, since the artisan who makes it gently shakes the mould while the sheet is being drained and formed. Because of the undefined pattern of the fibers, handmade paper has the most overall strength.

A flat machine that simulates the action of the artisan by shaking mechanically was invented by Nicholas Robert in 1798. This machine was perfected by the Fourdrinier brothers of England in 1803.

▶ In one mechanical system for making sheets of paper, the right amount of pulp is removed from the vat's interior with a strainer fitted to a rotating drum. This method expedites production. Small dried leaves have been added to this vat to make a run of decorative paper.

▲ ▲ When a piece of commercially-made paper is torn in the direction of the fibers, it results in a clean, continuous rip. If it is torn in a direction opposite the fibers, the rip is uneven.

NOTE THE FOLLOWING FEATURES	
In the Direction of the Fibers	**In the Direction Opposite the Fibers**
• The paper tears evenly • It folds easily • It's stronger • It shrinks less in drying • It takes print better	• The paper tears unevenly, in a jagged line • It's difficult to fold—requires more effort • It's not as strong • It shrinks more • It doesn't take print as well

Paper Strength

How paper is made effects its resulting strength, and this quality determines how it can be used. For instance, drawing paper needs a strong surface so that it isn't punctured by the point of a pencil. If the paper is going to be used to make envelopes, it has to be strong enough to keep it from coming apart or fraying when folded.

The following factors determine the strength of a sheet of paper:

- The type of fiber used. Remember that every fiber has a different degree of strength.
- The degree of refinement of the fibers. The more refined the fibers are, the less fraying they exhibit, and the better they join together.
- The weight of the paper. Cardboard will always be stronger than cigarette paper, for instance.

- The pressing of the paper. If the paper is not pressed, its strength is reduced because the fibers are not squeezed together tightly.
- The sizing, which makes it resistant to moisture and liquids. A paper without sizing is like blotting paper.

In a subsequent chapter on paper fibers, you'll learn about the strength of the most common fibers and how they are best refined to fit the paper's ultimate purpose.

An analysis of a sheet's strength and resistance is done in the following ways: stretching the sheet from its edges to find out its resistance to tension and its traction, folding it repeatedly to observe how it holds up, and tearing it to see how easily it rips.

▲ Refining in a Hollander beater allows the papermaker to control the pressure on the fibers. This factor, along with the length of the treatment, has a decisive effect on the paper's strength.

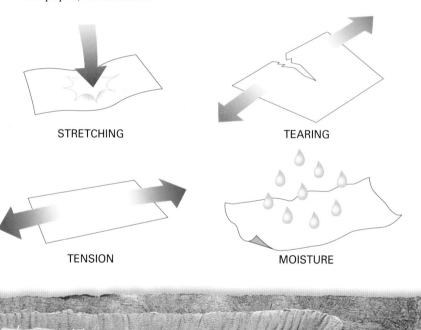

STRETCHING

TEARING

FOLDING

TENSION

MOISTURE

Paper's resistance to liquids is important when using it for applications such as binding and art media. For engraving and photography, the paper has to be completely submerged, and glue is added to this paper to give it the degree of resistance it needs.

The papermaker must control all of the factors—fibers, weight, pressing, and sizing—to create papers that are thin or thick, stiff or soft, and absorbent or water resistant.

When a piece of paper is preserved under ideal conditions, it becomes even stronger with the passage of time, like old wood that has been properly dried. (Keep in mind that it's made of the same material.) An old saying refers to "old paper and new ink," because it's well known that paper kept properly improves with age, like fine wines.

◄ The strength of the paper depends in part on the type of fiber used. Some plants have very strong fibers.

Heavyweight paper is thicker, and, thus, heavier, than light ones. You've probably judged the thickness of paper by touching it and feeling how thick or thin it is, whether it is letter paper, card stock, cardboard, or another kind of paper.

Paper is measured by square meters or yards. (A common 90-gram paper, for instance, indicates that a square meter of this paper weighs 90 grams.)

Making paper begins by weighing the dry pulp prior to refining it. The weight of the pulp directly determines how much paper can be produced. The paper ends up weighing the same amount as the pulp put into it. The amount produces a certain number of sheets of a particular size and weight.

HOW TO CALCULATE HOW MUCH PULP IS NEEDED TO MAKE A PRE-DETERMINED NUMBER OF SHEETS

Making paper begins by weighing the dry pulp prior to refining it. The weight of the pulp directly determines how much paper can be produced. The paper ends up weighing the same amount as the pulp put into it. The amount produces a certain number of sheets of a particular size and weight.

The following steps show how to calculate the quantity of pulp that needs to be refined to produce 10 thick pieces of paper measuring 20 x 26 inches (50 x 65 cm) each. We'll base it on the usual weight of this card stock, which is 350 grams (12.25 oz).

1. First, calculate the surface area of one card by multiplying one dimension by the other (20 x 26 in = 520 in^2/50 x 65 cm = 3,250 cm^2).

2. Multiply the surface area of a single card by the number of cards to be made to find out the total surface area (520 in^2 x 10 = 5200 in^2 total/3,250 cm^2 x 10 = 32,500 cm^2 total).

3. In order to find out how many yd^2/m^2 there are in 10 cards, divide the total surface area by the area in in^2/cm^2 of one yd^2/m^2 (5200/1296 = 4 yd^2/32,500/10,000 = 3.25 m^2).

4. Finally, in order to calculate the weight of these cards, multiply the total yd^2/m^2 by the weight of one m^2 (4 yd^2 x 350 = 1400/3.25 m^2 x 350 = 1,137).

So, in this case, the quantity of pulp that needs to be refined to produce 10 thick pieces of paper measuring 20 x 26 inches (50 x 65 cm) each is slightly more than 2 pounds or one kilogram of pulp (precisely: 2.5 lb/1.14 kg). In practice, it's a good idea to mix a little extra pulp to make up for shrinkage in the vat. In the present case, it's a good idea to mix up about 3 pounds (1.25 kg) of pulp.

CALCULATING THE AMOUNT OF PULP NEEDED FOR A SINGLE SHEET

Here's the procedure to follow for determining the amount of pulp needed or the weight of one sheet:

1. Calculate the surface area of the sheet by multiplying the length by the width.

2. Next, weigh it on a gram scale, and record the weight. (If a single sheet is too small or light to move the scale, weigh 10 sheets and divide the reading by 10.)

3. Next, apply the rule of three: If the surface area calculated in step 1 weighs the same amount calculated in step 2, how much will a square yd/m of this same paper weigh?

Let's look at an example: we have one sheet of thick paper that measures 20 x 30 centimeters and weighs 12 grams (remember that a square meter measures 100 x 100 cm =10,000 cm^2); then we apply the rule of three:

Surface area: 600 cm^2
Weight: 12 g
600/12
10,000/x
x = (12 x 10,000): 600 =120,000: 600 = 200

The paper in question has a weight of 200 grams (7 oz). It's at the limit that separates thick paper from card stock.

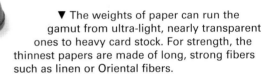

▼ A kitchen scale works for weighing dry paper.

▼ The weights of paper can run the gamut from ultra-light, nearly transparent ones to heavy card stock. For strength, the thinnest papers are made of long, strong fibers such as linen or Oriental fibers.

Paper, Card Stock, or Cardboard?

Any thick paper that weighs above 180 to 200 grams (6.3 to 7 oz) is considered to be card stock. Cardboard is paper that weighs over 350 grams (12.25 oz). A wide range of papers ranging from very fine to card stock also exist. Differences in transparency and refinement of these papers are very evident. Weight differences between card stocks are less obvious because they fall into a narrower weight range. Like paper, cardboard also has a range of weights from light to extremely thick and heavy.

▲ Handmade cards drying naturally at Ca l'Oliver in Sant Quinti de Mediona (Cataluña, Spain). The features that make these sheets much more attractive to artists than the common card stock available in the marketplace are the fibers, weight, format, handmade finish, and deckle edges.

◀ Card stock is stronger than paper because it's denser. It's a good choice to use for applications that require a certain degree of rigidity.

Cardboard

Making cardboard is as specialized as papermaking. In the old days, cardboard was also made in mills using different materials than paper. Recycled papers, ropes, and fiber sandals were often used to make this material.

According to a passage in M. de la Lande's *The Art of Making Paper* (*Arte de hacer el papel*) from 1768, banned books were often recycled into cardboard. By shredding and ripping them up, the government's prohibited books couldn't be pieced back together.

Today, recycled materials are still the main component of cardboard. Depending on the intended use of the cardboard, recycled newspaper print and lesser-quality mechanical pulps with strong wood fibers are used.

The following are the three most common ways of producing cardboard:
- Make very thick sheets all at once, using appropriate moulds and a larger amount of pulp;
- Form the sheet by adding a layer of pulp at a time while it is still moist, so that the layers adhere to each other;
- Glue sheets together to form a thicker layered sheet.

▲ Cardboard can be very strong, even stronger than wood on a weight-to-weight comparison. Very thick cardboard, which also contains hard fibers such as kraft pulp, can be used to make large items such as furniture and packaging.

◀ These old industrial cardboard samples illustrate the tremendous variety of cardboard applications that existed before plastics were made.

One characteristic of handmade paper is its unique tactile quality, especially when the surface is left natural so that it retains its air-dry finish. Other paper characteristics have evolved throughout history. For instance, the early Chinese polished the surface of their papers using agates to produce a softer feel to the paper, making it ideally suited for writing.

Rustic Paper

"Rustic" is a term used to describe handmade paper that isn't pressed after drying. Consequently, the surface is uniformly rough. Coarse woolen cloth is used to make this sort of paper, lending it a texture similar to fabric. Because it doesn't even get an initial pressing while wet, and is allowed to air dry slowly, it has an exaggerated, rough texture—even though it is soft to the touch and relatively smooth. This paper also goes by the names rough, hard grain, and others.

Vellum

As we discussed in the section on parchment, vellum is a type of fine parchment made from the softest hides of the highest quality. (This name also applies to heavy, off-white, fine papers that resemble parchment.) When vellum is lighted from behind, it looks completely smooth and uniform. Most vellum papers have a satin finish.

Consequently, these papers are still made with uniform mesh moulds. The deckle for vellum was invented in the eighteenth century by John Barkerville who wanted to make a paper as close to parchment as possible. He used very fine copper mesh to make the deckle.

Satin-finish Paper

This type of paper is dried before being passed through the press a second time to eliminate wrinkles and create a more stable surface. This process also modifies the tactile quality of the paper so that it is satiny.

For many centuries, the smooth surface of this paper was achieved by pounding it with a mallet or burnishing the sheets with highly polished stones or pieces of wood. Today, steel presses, screw presses, or polishers are used. Applied heat contributes to this type of finish. If only a small quantity of paper is involved, even a clothes iron can be used to smooth the surface.

This type of paper is also called by other names, depending on where it is made: matte, eggshell, medium, regular, light grain, and so forth.

 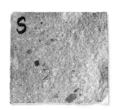

▲ A paper that isn't pressed when dry retains a natural roughness, especially if the fiber is very coarse (not highly refined). When pressed, it becomes smoother, making it possible to write or draw more clearly.

◄ A very rustic finish on handmade paper caused by taking pulp directly from the vat and allowing it to dry so that the fibers contract to form a rough surface

Many artists and preservationists prefer to use "acid-free" paper that lasts. The pH is a measure of acidity or alkalinity in paper and other materials. If the pH is at 0, it is acidic, and as it approaches a value of 14, it is alkaline. A neutral pH is seven. It's preferable for paper to approach this median value to assure its longevity. Although many papers have survived for centuries without this pH, it's usual for them to become yellow and brittle as the years go by.

Other factors that influence the permanence of paper are the percentage of cellulose in it and storage conditions. If the paper is made of 100% cellulose, it's more likely to stay white and free of oxidation, as is the case with most cotton and linen papers. In contrast, mechanical pulps using wood fibers deteriorate a lot over time because these pulps contain not only cellulose, but also other components from trees (lignin, lipids, minerals, and so forth). For conservation purposes, paper should be kept in a dry environment away from air currents, since humidity weakens it tremendously and causes spotting. The ideal environment is a temperature of 67°F (18°C) with 65% humidity.

▼ Various finishes on handmade papers are created by using rough fabrics or other materials and treating the paper in certain ways.

Vergé Paper

There are several interpretations of the meaning of the term "vergé paper." The French expression *papier vergé* means "ribbed paper." The English adapted the French term *vergé wire*, meaning a mesh made of "ribbed wire." There is also an interpretation that links the term to the marks that can be produced on the paper by striking it with a flexible *verge*, or *rod* in French.

In general, the term designates paper that is made using a type of mould with a configuration of thin, woven rods and wire that leave impressions on the paper. The rods can be made of bronze, brass, bamboo, or any other material that is water-resistant. When a sheet of vergé paper is held up to light, marks from the weave can be clearly seen. The thickest and most numerous lines are the marks made by the rods, and the finest, perpendicularly-oriented ones are the marks from the wire that weaves the rods together.

Vergé paper has less strength lengthwise than widthwise, because this area contains less fiber. The width of the paper appears lighter when backlit, and it tends to tear along these lines.

► Not all vergé papers are the same. They vary according to the thickness of the rods, the tightness of the weave, and even the age of the mould. With use and age of the mould, the lines become more irregular, lending the paper more character.

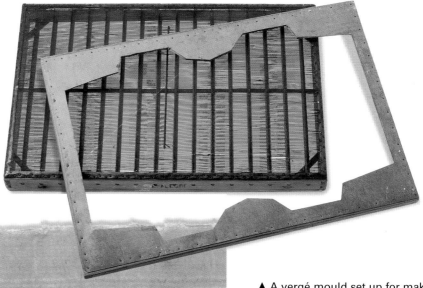

▲ A vergé mould set up for making two envelopes

◄ Handmade vergé paper from Japan

Coated Paper

To fit the needs of mechanical printing, manufacturers make coated papers that are very stable, opaque, and luminous. In the process of making this paper, a fine layer of special sizing (coating) is added on one or both sides to fill in all the pores and even out the surface. Sizing optimizes the paper's use for printing books and magazines. These papers are often somewhat glossy. Later in the book, the coatings for this type of paper will be discussed in more detail.

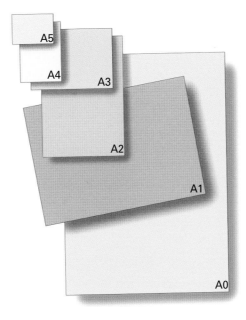

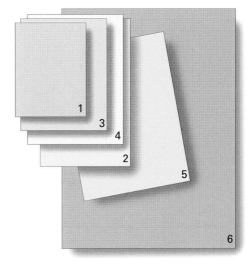

◀ ▲ Some of the oldest paper sizes and formats continue to be used today because of their versatility for art and typography.

Paper is made in a wide range of sizes, and the size of handmade paper can be varied simply by changing the dimensions of the moulds. Before paper was manufactured in a continuos roll, different papers in various sizes, colors, finishes, and so forth were milled to accommodate different uses. For example, different papers were used for writing love letters as opposed to composing a letter to a distant relative, as well as for recording a notary act as opposed to an accounting procedure. Wrapping lace or velvet also required their own distinct papers. Consequently, many mills specialized in one or several types of paper that were exported to other countries.

In the eighteenth century treatise, *The Art of Making Paper* (*Arte de hacer el papel*) by M. de la Lande, there is an extensive inventory of papers manufactured in France and Holland; all with different dimensions, compositions, and weights. Because of this broad array of sizes available in the marketplace, the paper contained a watermark that identified it by sight without having to describe the dimensions. The majority of these papers are no longer manufactured, but some are still made today because of their size and acceptance in book publishing and the arts. Today, dimensions are stipulated. The following lists specify some of the papers still in use:

Paper Names:
1. Couronne . . . . 14.5 x 18.5 in (37 x 47 cm)
2. Double Couronne . . . . . 18.5 x 29 in (47 x 74 cm)
3. Coquille . . . . 17.75 x 22 in (45 x 56 cm)
4. Double Raisin . . . . 19.7 x 25.6 in (50 x 65 cm)
5. Jesus . . . . 22 x 30 in (56 x 76 cm)
6. Double Colombier . . . . 35.5 x 47.25 in (90 x 120cm)

Today, there are many systems for indicating the size of papers. In Europe, sizes are shown by "DIN formats," as listed below and shown in the diagram at the top of the page:

A0 . . . . 33 x 46.8 inches (841 x 1189 mm)
A1 . . . . 23.4 x 33 inches (594 x 841 mm)
A2 . . . . 16.5 x 23.4 inches (420 x 594 mm)
A3 . . . . 11.7 x 16.5 inches (297 x 420 mm)
A4 . . . . 8.25 x 11.7 inches (210 x 297 mm)
A5 . . . . 5.8 x 8.25 inches (148 x 210 mm)

HISTORICAL UNITS OF MEASUREMENT

Paper was traditionally measured in units such as a *bale*, *ream*, *signature*, and *sheet*. These units were standardized around A.D.1200. Today, some of these measurements are still used, but vary according to printing requirements. The word ream comes from the Arabic word *rizma*. This unit traditionally equaled 500 sheets. The term is used throughout the world: *resma* in Spanish, *risma* in Italian, *rame* in French, *Ries* in German, *ries* in Danish, and so forth. Historical standardized measurements for paper are listed below, although some vary today:
- A *signature* consisted of five sheets
- A *ream* was 25 quires (500 sheets)
- A *quire* contained five signatures (25 sheets)
- A *bale* was ten reams (5000 sheets)

▼ Lluis Morera and Toni Capellades making a large-format sheet by removing it from the vat and gently agitating the mould to uniformly spread out the paste. Large-format sheets can be made easily by two people.

Sizing and Coatings

The hydrophilic quality of cellulose means that every sheet of paper is a potential sponge, ready to absorb moisture and lose its shape. To avoid this and make the paper more stable and strong when exposed to water, sizing glues are added to it. Without sizing, ink would run on the paper, and the strongest and driest paper in the world would disintegrate if it came into contact with water.

The origins of sizing for paper are as old as the invention of paper. In fact, sizing was explored before paper was invented in China. (You may recall that T'sai-Lun experimented with various glues on silk fabrics to create a primer that makes it possible to write clearly on them.) Different kinds of glues have been used on paper throughout history, depending on the location of the mill and the technological advances in the region.

In the Orient, the first sizing contained algae, and wheat starch was used in the Arab world. Beginning in the thirteenth century, animal glues made from the wastes of tanneries were used. In the twentieth century, synthetic paper sizing was formulated to provide moisture resistance, without any adhesive function.

▼ Paper without sizing becomes blotting paper.

▼ Sizing makes paper waterproof.

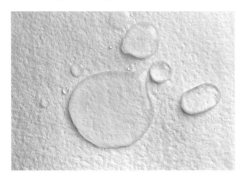

▲ Engraving 12 from *The Art of Making Paper* (*Arte de hacer el papel*) by M. de la Lande, 1768

Sizing can be distinguished according to its makeup, whether derived from animals, vegetables, or created synthetically. In the chapter on materials and tools, we'll devote some space to this topic. Next, we'll discuss the two main methods of applying the sizing: external (superficial) or internal (in the pulp).

External Application

External application of sizing is the oldest method, and it was used exclusively for making writing papers. In this application, fully dried sheets in their final form were submerged in a solution that penetrated the paper to a certain degree, providing greater strength and impermeability. The baths might contain animal sizing (usually made from the wastes produced by butcher shops and tanneries) or plant sizing (made from rice, corn, wheat or tapioca starch, algae, plant juices, or rosin). After the bath, the coated paper was dried and pressed again.

These types of sizing were applied until the end of the eighteenth century. From that time on, internal application of sizing was practiced in Germany. A comparison of papers that were sized externally and internally reveals a considerable difference in appearance and feel. The paper that was coated externally is stiffer and harder than the internally sized paper. If animal sizing was used, the paper was also glossier.

Internal Application

The invention of the Hollander beater in the early nineteenth century, a machine designed to facilitate the refining of paper fibers, led to experimenting with internal sizing. The beater considerably reduced the time needed for refining paper, changing it from a process that took days to one that required just a few hours.

If sizing was added during refining, the paper became stronger and more water resistant, since the sizing was more effectively integrated. Unlike external application of sizing, which didn't fill the pores of the paper, this type of sizing did, providing a better surface on which to write.

Today, internal application is used in handmade paper. Sizing is added to the vat during refining or at the stage when the pulp is ready to be made into sheets.

▲ The sizing area of the Moli Paperer de Capellades Museum in Barcelona, Spain. Note the kettle in which the sizing was prepared, the copper vat where the sizing took place, and the small press used to encourage the penetration of sizing in the sheets. The excess glue from the pressing was returned to the copper vat.

Coatings

Coatings are fillers added to the cellulose pulp to improve certain features of the paper, including the way it takes ink, its strength (against tension, breaking, folding, or moisture), and keeping added materials from separating.

Fillers change the weight of the paper and lower the cost of it by reducing the amount of cellulose fiber used (from five to 40% in a sheet). Naturally, the durability of the paper diminishes along with the purity of its composition.

Fillers can originate from the following sources:

- Glues made from animal products
- Starches from rice, wheat or tapioca, resins and gums, etc.
- Minerals such as calcium carbonate, kaolin, talc, titanium dioxide, etc.

Mineral fillers are the most common and are frequently used in making coated papers. They are intended to cover empty spaces between the fibers, and improve the surface of the paper (whose normal appearance is rough with creases, ridges, and hollows). Since these particles are much smaller than the fibers, they're very effective for filling in the empty spots. Papers containing fillers take ink well because they absorb it quickly, unlike pure cellulose. This is an important factor in printing, because it prevents smudges.

Like glues, fillers can also be applied externally or internally, depending on their final use. Adding them internally provides more stability.

▼ Jordi Catafal, *Cookbook* (*Libre de cuina*) from Scala Dei.

This bibliophile's book contains ten engravings by the author. The text is from the book of kitchen recipes from the former monastery of Scala Dei (Tarragona, Spain). The pages are scented with an essence developed by Rossend Mateu and Josep Carrera, fragrance distillers, based on the scents of regional plants, to evoke olfactory sensations from the monastic cuisine.

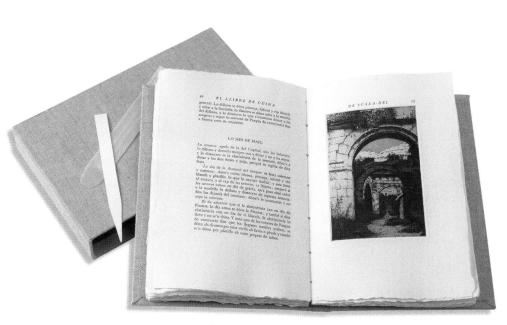

◀ Samples of decorative papers incorporating added plant materials

▼ Newspaper print is one of the lowest quality coated papers. The filler added to this paper effectively lowers its cost by reducing the amount of cellulose in each page. The filler also facilitates mechanical printing.

Added Materials

The fibrous structure of paper makes it possible to introduce foreign materials and integrate them perfectly. In the past, no material of any kind was added to the paper unless it accomplished some practical function; only fillers were added. But as paper was used in other ways, such as for the purposes of art and decoration, people experimented with this process with some amazing results.

The aesthetic quality of paper enhanced with added materials can be wonderful, but the drawback is that the additives do nothing to enhance preservation. We have no idea, for instance, about the future condition of papers that contain flowers or algae, since most of the samples are not over 100 years old. It's highly probable that pages containing additions such as these will have serious preservation problems long before they're anticipated. Keep in mind that each added material is a foreign body in the sheet.

▲ Caroline Greenwald, *Flight Wings*, 1972, private collection.
This delicate artist's book is made of *amate* paper, silver paper, manila papers, tengujo, and *gampi* with white deer hair added.

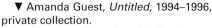

▼ Amanda Guest, *Untitled*, 1994–1996, private collection.
In this piece, the artist added nails to handmade paper to create tension during the drying process. By interfering with the uniformity of the paper, the artist accomplished the result that she wanted.

Practical Considerations

There are a number of things to keep in mind when adding foreign materials to paper:

- The materials shouldn't float in liquid because they will be too difficult to integrate into the sheet (e.g., feathers, straw, cork, wrapping papers).
- The materials also shouldn't settle to the bottom of the vat (e.g., stones, metals). If these materials are added to paper, they have to be sandwiched between layers.
- It's a good idea to soak the materials before adding them to the pulp—in some cases for several hours.
- The materials that integrate most effectively with pulp are flexible; the least desirable ones are stiff, dry, and brittle. For instance, fresh plants are a better choice than dried ones.
- Always avoid using materials that decompose and rot quickly, such as fresh foods.

How to Incorporate Foreign Materials into Paper

There are four general techniques for adding foreign materials to paper. The one that you choose depends on the composition and type of material being added. The final result can be left to chance, or you can control the arrangement of the elements as much as possible for compositional purposes.

1. Adding materials in the vat: The materials remain in aqueous suspension, like the pulp, and they integrate very well.

2. Adding materials in the mould: This technique is used when the sheet is being formed. It's helpful to have another person lend a hand. Also, use a mould with small pores and a raised deckle in order to prolong the draining and provide more time to do the work. If you see that the material is not fully integrated into the surrounding fibers, you can add a little bit of diluted fiber on top after draining and before couching.

3. Adding materials to the felt: Before couching the sheet, add a thin layer of highly diluted pulp to the felt. Then carefully place the elements on the pulp.

4. Sandwiching materials between two paper sheets: Use sheets as thin as possible to reveal as much as possible of the item between them. Place the first sheet on the felt followed by the added material. Add the top sheet, then press the two sheets together.

Textile Inclusions

Textile inclusions integrate well into paper, and there are few preservation problems because the fibers used to make textiles are the same as those used to make paper. (Even though these fibers haven't been subjected to a previous chemical treatment). In addition, textile fibers add strength to the paper, since they act as a core and create a better filament structure.

Very attractive results can be produced by using jute, hemp, pita, and all sorts of rope fibers. Paper can be made using all these fibers if they're initially boiled with soda and refined. Lots of antique papers have bases of these sorts of fibers, derived from wastes from making sandals, baskets, seats, hats, bags, ropes, and so forth. Paper made from these fibers is rather coarse and rustic, but it's strong and durable.

Another very attractive element that can be added to paper is colored thread, whether linen, cotton, silk, or even nylon. Threaded paper is very elegant and can be used for many applications. Pieces of fabric, such as nets, fabric remnants, or braided threads can also be incorporated into paper with their warp and weave intact. You can produce nicely integrated effects by carefully placing these inclusions on the sheet when it is wet.

▲ Many fibers used for making sandals, baskets, and other textiles are very well suited to making textile-like papers.

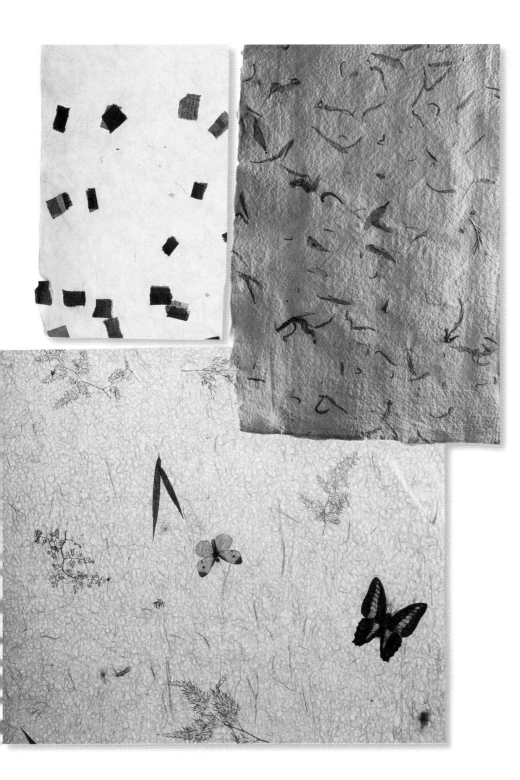

▲ Threads, string, and fibers of varying thickness increase the strength of paper and make it attractive.

◄ Decorative papers can also be purchased. They are often used as gift wrapping papers, stationery, or bookbinding materials.

Flowers & Plants

Paper containing flowers is probably the most popular kind of paper with added materials. The flowers are embedded in very thin paper, making the shapes and colors of the flowers clearly visible. Normally, only the flower's petals are used since the other parts contain substances that can stain the paper. In addition, the bulk of the rest of the plant reduces the paper's stability and contributes to tearing.

Through experimentation, we've discovered that some flowers are better to incorporate into paper than others. Some have petals that end up rotting or staining the paper. Some lose their color quickly—just minutes after they are put into the vat, while others retain their color for many years as if still fresh. There are still others that gradually dry and fall out.

A number of factors influence how well the flower works in the paper: the type, when it was picked (there are many hypotheses about the influence of the seasons, the lunar cycle, or the time of day), how it was preserved before being added, and the temperature (the best petals are those obtained in cold temperatures, and it's helpful to freeze the petals before putting them into the fibrous suspension).

We know for certain that small or elongated petals work better than larger ones that have a bigger surface area, and dried flowers almost never work well because they end up staining the sheet a dark color.

The petals of flowers such as daisies or chrysanthemums are good bets. It's a good idea to moisten the petals before putting them in the vat because it softens them and slightly dissolves their waxy coating.

Petals aren't the only materials that can be used to make natural papers. Other parts of plants, especially leaves and bark, can be incorporated into paper. Ferns, stalks of grain, and asparagus leaves work well for example. One interesting, classic method uses partially-ground onion skin. If the skins are boiled beforehand, they integrate better, and the water used for boiling can also serve as a natural dye for the paper.

▲ The calendula is one of the best flower choices for use in paper.

▲ Grain parts, such as rice hulls and straw can be added to paper. Because many grains are water-resistant, particularly straw, they require boiling before adding them.

Aquatic Plants and Lichens

Algae and many other aquatic plants are compatible with the aqueous nature of paper pulp because they exist in a watery environment. There are many aquatic plants in the world and a tremendous number of possibilities from which to choose. All of these plants integrate well with the paper pulp; but, when they dry, they react differently. Not all aquatic plants produce good results after drying.

For instance, some become stiff, hard, or brittle when they dry, while others retain their flexibility. Before adding any aquatic plants to paper, it's essential that they be cleaned thoroughly to remove microorganisms and sand.

◄ Papers containing leaves and flowers that have been added with great care

Freshwater weeds as well as seaweed can be used as added materials. Aquatic weeds are usually found where there is running water and live organisms. They can also be purchased at markets where they are sold as food, but many of them are quite hard when dry. People have attempted to make paper entirely from water plants, but a tremendous amount of fiber is needed to make a single sheet, since it weighs very little after drying.

Lichens can also be added to paper for nice results, and most retain their original color and flexibility for years. They also combine nicely with aquatic plants because of their shape and color.

▲ Lichens gathered from trees in various woodlands

▲ Aquatic weeds picked up in the Sella River (Asturias, Spain) during a canoe trip. These materials can be gathered and saved for special projects later.

Materials with Volume

If incorporating materials with volume, or depth, into paper, the fibers will contract around them during drying to create interesting effects. If thicker materials are covered with recently made paper, the paper will air dry and conform to the shapes. In these cases, the sheet is intentionally left unpressed to prevent damaging the materials. Creative projects using this principle can result in some very interesting relief textures. These sorts of projects are different from paper with inclusions.

Metals

Because of its weight, metal doesn't integrate well with paper; however, it can be added to the vat in the form of powder or glitter to produce some interesting decorative papers. Thin, flexible metal such as copper filaments, mesh, and screening can be sandwiched between two sheets of paper. But, keep in mind that metals oxidize in water. You can choose to integrate this effect into the paper's aesthetic, or try to avoid it by speeding up the drying process with fans or a heat source.

◄ Two different types of aquatic weeds. There is a tremendous variety of seaweed in the ocean, but not all of them respond well to drying. Avoid fleshy ones and those that harden after drying.

► Confetti gives paper a festive appearance.

Animal and Insect Materials

Materials such as hair, scales, insect wings, and the skin of fish and reptiles are sometimes incorporated into paper for beautiful effects. However, these materials don't integrate easily with the paper since they contain a high level of fat that causes water to run off their surfaces. Because of this factor, the papers must be made in a mould or on a felt rather than in a vat.

▲ A group of butterfly wings discovered at the edge of a rural road. While it's difficult to incorporate insect wings into paper, the results can be spectacular.

Other Materials

Decorative papers can be made by adding remnants of other papers to the pulp—magazine pages, road maps, or confetti for example. Cellophane paper, aluminum foil, and candy wrappings can also contribute color and interest to paper.

Spices such as curry, saffron, oregano, sage, and thyme can be added to give the paper a natural aroma that lasts a long time but eventually disappears. Other more-fragrant perfumes for paper can be applied after the paper is dry using a bath or aerosol. Before using perfume, check its acidity to avoid yellowing or staining over time.

Paper Color

Another characteristic that defines a sheet of paper is its color. Historically, the color indicated the purpose of the paper—for example, deep blue paper was used for bundling lace, grayish blue paper wrapped packets of sugar and candles, and yellow paper was used for sketching. In the past, the spectrum of paper colors was very limited. Today, paper colors are almost unlimited, from pastels and bright colors to earthy neutrals and grays.

▶ Paper made using various species of woodland bushes. The dark green color is due to the lignite and extracts that were very difficult to remove in the cooking process.

The Natural Color of Fibers

Different fibers have their own distinct colors. When the fiber comes from a plant, the color remains the same after the chemical process that separates the cellulose from the other materials. The color of plant pulps is usually a dark yellow or ivory. Kraft pulp has the most familiar color; it comes from coniferous trees whose brown color is typical of wrapping papers, cardboard boxes, and paper bags. In contrast, rag pulps tend to be exclusively white. (This type of paper isn't made very often today.)

Pulps made from other recycled materials, such as rope, fishermen's nets, fiber sandals, and baskets, have a darker color in the realm of ochre yellow or gray, depending on their composition. In eighteenth-century Europe, soot was added to these pulps to accentuate their dark color. Because of its inferior quality, this paper was used for wrapping white textiles. The dark paper made them appear brighter and more attractive to the buyer.

Papers made from recycled printing paper is gray because of the printer's ink, which is very difficult to remove. This gray characterizes recycled paper that is often tinted with blue, pink, green, and other colors.

▶ Paper made from straw

▲ Corn paper showing the typical color of the cobs

▼ *Kozo* fiber. This oriental plant is used in making "Japan" paper. It is naturally light in color when it's gathered from the bush, as you can see in the bundle. Then it's moistened so that it can be stretched, peeled, and boiled with sodium carbonate. The resulting natural color is very white and doesn't need any final bleaching.

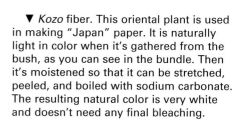

◀ Banana paper made using the leaves and bark from banana trees

Bleaching

Long before the invention of paper in China, people used white surfaces to write on. The whitest parchments were considered to be the best, and vellum and papyrus were light and luminous. Logically, the same qualities were sought in paper. In the towns of the Far East, light fibers that were easy to bleach were readily available. Consequently, Oriental papers have always been known for their whiteness.

Kozo fiber, for example, is naturally light in color and becomes even lighter when the impurities are peeled off the bark before boiling it with sodium carbonate. If it is exposed to sunlight, the whiteness increases even more. The *mitsumata* fiber is a luminous white that shines like silver.

Until the twentieth century, moist fibers were bleached in the sun to lighten them. The natural pigments lost their color, dissolved, or were destroyed. The exposure to the sun was alternated with successive soakings and boiling in alkaline solutions containing wood ash, and applications of lactic acid.

This treatment was used on cotton and linen fibers, which didn't require a lot of bleaching. For darker fibers, a lot more bleaching power was needed. Due to the high cost of this bleaching process, the dark color was simply accepted.

In 1774 the Swedish chemist Carl Scheele discovered chlorine and tested its whitening power on plant fibers. Shortly thereafter the Scotsman Charles Tenant patented a whitening powder made from chlorine and diluted calcium carbonate. This patent made it possible to apply the reagent in solid form.

When chlorine is applied to fibers in a bleaching vat, 80 to 90% of the reagent is consumed in the first 15 minutes. The ultimate duration of the reaction depends on the quantity of reagent, the temperature, and the consistency of the fibers. Most pulps are chlorinated at a low concentration (from two to five percent), at room temperature (68 to 86°F/20 to 30°C), and for a time that varies between 45 and 60 minutes. A thorough washing of the pulps is required after bleaching. The process can be repeated if more whitening is needed, but care must be taken to avoid weakening the fibers.

Today bleaching is done with chlorine or its derivatives—ozone, oxygen, and oxygenated water.

▶ Natural colors as well as whites are very desirable in the commercial market. Here, they have been incorporated into the design of these notebooks to produce a simple yet beautiful product.

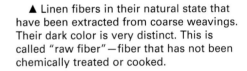

◀ In this Thai booklet of *sa* paper, it's easy to see the contrast between the natural color of the fiber in the covers and the brighter color of the bleached pages inside.

▲ Linen fibers in their natural state that have been extracted from coarse weavings. Their dark color is very distinct. This is called "raw fiber"—fiber that has not been chemically treated or cooked.

▼ Linen papers made from raw, bleached fibers. Raw linen has a dark color, but it can be lightened to produce a bright white.

Old Ways of Coloring Paper

As early as the eleventh century, the Arabs were using saffron to dye paper yellow and sycamore sap to dye it brown and make it appear ancient. In the seventeenth and eighteenth centuries, the Dutch were dying paper with natural substances to tint it light blue.

Paper was traditionally colored with natural materials such as saffron, sycamore, soot, indigo, tea, and onion. The pulp could be tinted while the paper was being made, or the color applied externally after it was dried. To create brighter colors, white pulps were tinted; but pulps with a darker natural color were used to create more subdued tones.

Fibers are tinted through capillary action that occurs when the colors enter them through their pores and central duct. Then a series of reactions happens between the colorings and the fibers that colors the pulp. Heat aids this process and was used in traditional coloring operations.

Not all fibers absorb color in the same way, and, as a result, variegated papers were made for centuries by mixing different colored pulps.

▲ In order to color these envelopes, the pulp was tinted with solid pigments during the refining process, resulting in a spectrum of pastel colors.

► Samples of colored papers. It's a good idea to keep a small sample of every type of paper made for the purpose of choosing colors and planning future projects.

▼ A sample of bright colors that are popular on the commercial market

Coloring Today

Today, pulp is colored during refining, since this helps to distribute color evenly throughout the whole batch. If coloring isn't done at this stage because it impacts the pulp's production, the fibers can be colored in the vat after they're refined. In this case, the pigment still needs to be thoroughly mixed with the fibers.

In theory, any coloring agent that's soluble in water can be used, but it's best to use special dyes for papers because their molecular structure produces a better tint. In chapter four on materials and tools, we'll discuss the topic of pigments more fully.

Creativity and Color

The use of color in pulp is, of course, not limited to commercial products. In the artistic realm, papers are often designed as a ground for different kinds of expression. For instance, late-twentieth-century British artist David Hockney used colored paper pulp to make a series of pieces that relate to his swimming pool paintings called the *Paper Pool* pieces.

Many other contemporary visual artists have used paper creatively, some incorporating relief and collage. The well-known, hand papermaking studio Dieu Donné Papermill, Inc., located in the Soho district of New York City, was founded in 1976 for the purpose of reinventing and adapting age-old techniques of hand paper-making to contemporary art.

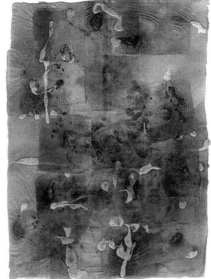

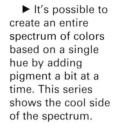

▶ It's possible to create an entire spectrum of colors based on a single hue by adding pigment a bit at a time. This series shows the cool side of the spectrum.

▲ Wen Yi Hou. *Life Unit*, 1996. Collage of handmade, dyed papers made from various fibers. Created at Dieu Donné Papermill (New York).

◀ Chuck Close, *Susan*, 1988. Colored pulp. This famous American artist created a series of pictures made from colored pulp in conjunction with the prestigious Wisconsin paper artist Joe Wilfer, who died in 1995.

▶ Papers colored in a warm spectrum

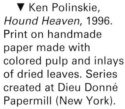

▼ Ken Polinskie, *Hound Heaven*, 1996. Print on handmade paper made with colored pulp and inlays of dried leaves. Series created at Dieu Donné Papermill (New York).

▼ Bill Weege, *Untitled*, 1981. Handmade paper with relief. Series created by paper artist Joe Wilfer.

Watermarks

Watermarks are designs in paper that are visible when the paper is back-lit. Usually, they are simple graphic designs or words with a clear identifying feature, like stonemasons' marks or signatures. Water helps to distribute the fibers that make up the mark.

A historical precedent for paper watermarks were decorative designs woven on a metallic mesh using metal wire, which will be discussed in more detail later in the book.

The History of Watermarks

The earliest paper mills didn't use watermarks to identify their products. The Italians were the first to use this type of mark. Fabriano, Italy, is credited with the invention, but the earliest preserved watermark is in a document from Bologna dated 1282.

Through the years, watermarks changed from being marks used to identify the craftsman to serving as an indication of the geographic location of the mill or quality, and especially the dimensions of the paper. They identified the paper at a simple glance.

Watermarks also served to commemorate important historical events, as they still do today. All of these uses have made watermarks more than simply an aesthetic element. They also serve as a practical and valuable asset for historians and researchers who are able to find out the date and origin of many documents.

▲ Watermark as it appears through paper in the Moli Paperer de Capellades Museum in Barcelona, Spain

◄ This delicate, portrait-like "sketch" created by Pablo Picasso for Jacqueline served to make the watermark she used on all of her stationery. The image is created using copper wire soldered and threaded through laid mesh.

Characteristics of a Watermark

Normally, watermarks are like line drawings that are made using metal wire threaded onto the screen or deckle of the mould. In the moulding process, the paper sheet retains less cellulose in the areas occupied by the wire. When the paper is held up to light, the difference in thickness is visible: the area is lighter due to the trace of the wire on the sheet.

Watermarks can also be as complex as chiaroscuro images made with a bas relief on a metallic mesh. The watermarks on paper currency are often tremendously complex portraits or landscapes with amazing spatial perspective. More pulp is deposited in the valleys of the bas relief than the high points, and these differences in thickness produce the delicate gradations. When lighted from behind, they appear highly representational.

Another important consideration is the location of the watermark in the mould. The mark is often placed and woven in so that it allows for making folds or cuts in the paper or is more visible in certain positions. Some historical watermarks, such as those for commemorative documents, were very large and were located on the central axis of the paper.

▲ Watermark with the signature of the Spanish poet Federico García Lorca (1898–1936). The artist's signature was incorporated into the paper on which engravings of his poems were printed so that every sheet was personalized.

▼ This watermark with the signature of the Spanish artist Joan Miró (1893–1983) was created for making the paper for a limited-edition book illustrated by the artist.

▼ Carter Hodgkin, *Gaussian*, 1991. Series by Dieu Donné Papermill (New York).
The poetic force of this work of art lies in the interplay of opacity and transparency created by the positive and negative of a watermark in the linen sheet.

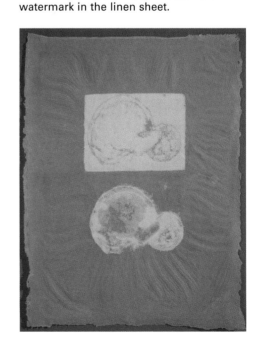

Typology

The first watermarks represented very simple geometrical figures such as crosses, numbers, and signs. Much more elaborate and artistic watermarks didn't appear until the beginning of the fourteenth century. These took the form of images such as hearts, shields, suns, birds, and religious symbols.

The number of watermarks made between the thirteenth and twentieth centuries is unknown. Investigations have been carried out in restricted geographic areas—one by the researcher Oriol Valls includes a detailed inventory of all the watermarks in the history of papermaking in Cataluña, Spain (Oriol Valls, *Paper and Watermarks in Cataluña*, The Papers Publication Society, Amsterdam, 1970). The use of watermarks is not limited to a single country or continent. The designs for watermarks quickly grew beyond the confines of standard types, making room for more creative ones.

In general, the most common designs fall into the following categories:

- Human figures, such as a head, bust, or hands
- Animals, such as snakes, lions, dolphins, eagles, bears, roosters, birds, dogs, ox heads, and unicorns
- Agricultural themes, such as tassels and grapes
- Military and heraldic symbols, such as helmets, arrows, armor, shields, fleur-de-lis, towers, and crowns
- Religious or astrological symbols, such as half moons, suns, stars, anchors, and crosses

▲ Several watermarks incorporating the sun as a symbolic design

◀ Mermaids, like unicorns, are mythological creatures that appear regularly in the iconography of watermarks.

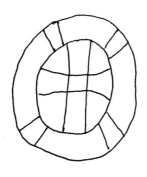  

▲ The Western watermark symbol par excellence is the cross.

 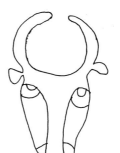

◀ Another recurring watermark theme is the ox's head.

▼ Images of elephants have often been used as watermarks to indicate strong papers. These two watermarks were created five centuries apart.

▼ Watermarks indicating the geographic location of the mill

Deckle Edges

In the case of commercially-made papers, the edges are usually cut to produce a precise format; however, it's more natural for the paper to have a ragged deckle edge because of its fibrous structure.

Until the nineteenth century, a frayed edge was considered to be a defect in the page, and it was cut off to provide a more regular edge. Very sharp blades were used to remove the irregular edges of the piled and pressed pages. If you look at the finish of page edges in books dating prior to the nineteenth century, you can often see the marks left by the blade. You can also see that they aren't cut at perfect right angles, since this process wasn't yet perfected.

During the nineteenth century, as commercially-manufactured paper became more widespread, there was more interest in handmade paper with deckle edges. When continuous-roll paper made from wood pulp took over most of the world's markets during the twentieth century, almost all the mills that were still making handmade paper closed their doors due to the competition. From the time of that radical change, deckle edges have been greatly respected and regarded as a symbol of a handmade product.

Once deckle edges became popular, a method for creating them was perfected. What happened naturally in the absence of excessive control and was once considered a defect turned into something desirable. From that time on, artisans chose their own distinctive edge for their papers, and this mark of the maker personalized their product.

▲ As this photo shows, not all deckle edges are the same, and each edge has certain distinctive qualities. A frayed edge can be the most unifying attribute in a section of similar sheets.

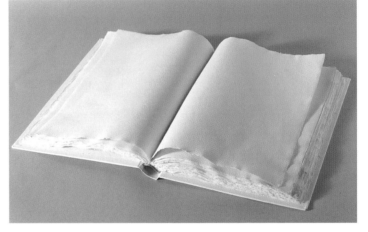

◄ Deckle edges provide a soft, special touch and warmth to books designed for personal use. The charm of this album lies in the beautiful irregularity of its edges.

► Historically, pages were trimmed on a bench where the deckle edges were cut off.

Forming Deckle Edges

Deckle edges are produced by two factors: the pressure applied to the sheets while in the press, and the limits imposed on the fibers as they are shaped in the mould.

The tool that is used for making paper is made up of two parts:

- The mould made up of a mesh frame that serves as a strainer and allows water to drain as the sheet as it is formed; and
- The deckle that secures the fibrous pulp at the edges and determines the size of the sheet while allowing water to drain through the mould.

When paper is made by hand, it's relatively easy to control the mould and deckle. The desired deckle edges can be made using bevels and shims in the deckle and insulators in the mould.

The idea is to create differences in thickness at the edges while the sheet is being formed and removed from the vat. To do this, fiber must be added that is slightly beyond the inside limit of the deckle. When the sheet is pressed, this small amount of fiber is forced beyond the inside dimensions of the deckle. This part of the sheet looks lighter, fibrous, and luminous when held up to a light.

The traditional way of assuring deckle edges involves using a deckle with a good bevel. The depth and angle of the bevel determine the nature of the deckle edge.

Another method involves adding tape to the underside of the deckle to lift it up slightly and allow a small amount of pulp to extend into the narrow space.

A final method involves making the inner dimensions of the deckle slightly larger than the mould's mesh (where drainage takes place).

▼ These illustrations show how the differently shaped frames can produce variations in the deckle edges. The first three shown are beveled. The last shown will always produce a small frayed edge, since no bevel is used.

1

2

3

4

◀ Ángels Arroyo, *Album*, 1996. Private collection.

Handmade paper, engraved and oxidized iron, parchment, and ink. This artist from the Invisible Ink Workshop in Barcelona, Spain, has maximized the diaphanous quality of the paper by contrasting the soft fraying of the pages' edges with the hard edges of the iron cover.

Imitation Deckle Edges

Imitation fraying can be created by carefully tearing handmade paper in an irregular fashion (without using a straightedge) and slowly increasing the length of the tear. A method that is commonly used on commercially produced papers is grazing the page edges with a saw after they've been carefully aligned and held in the press to prevent tearing or wrinkling.

▼ Deckle edges stand out because of differing thicknesses as well as the ripples they add to the edges of the page. Ripples form during the drying process, which starts at the edge of the sheets and progresses toward the interior, producing tension in the fibers.

Textures and Crimping

The cellulose fibers that make up paper are very malleable when hydrated. When they dry, they recover their natural stiffness and become hard. Because of this characteristic, it's possible to create different surfaces and textures by using different kinds of moulds or by manipulating the sheets during the process. For instance, if you use a flat, smooth mould, it will produce flat, smooth sheets of paper.

Nearly all artisans end up manipulating the textures of the papers they create, since it is the tactile quality that makes one surface differ from another. We've already looked at some of the most common surface textures of hand-made papers, such as vellum, rustic, laid, satin, sized, and so forth. However, let's also look at some experimental textural possibilities.

▲ The texture of this paper was created by allowing it to dry in the elements without pressing it. The small pockmarks are made by raindrops. (You can achieve a similar effect with a hose, but you have to be careful to avoid ruining the sheet.)

◄ These notebook covers were created by air drying hand-made cardboard and imprinting the forms of mollusks to suggest fossils embedded in a rocky surface.

▲ Samples of textured papers showing marks from sacks and window shades that give the paper a natural and human character.

Textures of Rough, Woven Cloths

While it's completely hydrated, paper can be textured by impressing it with another material. Under pressure, the paper adopts the texture with great precision. For example, if you place an old, stitched and torn cloth sack between the paper and the felt, the pressure of the press will imprint all the details on the sheet.

Many artisans use this idea, and keep a large number of textiles, such as sacks, cloth window shades, and interesting weavings on hand to use as felts for making papers with appealing textures. (The textiles that are used must be somewhat water-resistant so that they can be separated from the paper when they're removed from the press.)

◄ Sacks and window shades used for adding texture to paper

Dry Crimping and Texturing

Crimping or *goffering* means adding a significant relief to a sheet that's already been formed and dried. This effect can be a surface relief (which is more textural) or the imprint of a shape. In either case, force has to be used on the paper since it has already dried and won't readily adapt foreign textures and shapes.

To produce a relief once the paper is dry, it is passed through a press or a roller die along with the textured surface or item. The following tips help when making a relief of this sort:

- Use papers that have long, flexible fibers.
- Moisten the paper a bit so that it will adopt the shape or texture of the material.
- For making the relief, choose strong, stiff materials that can withstand the pressure of a run without breaking.
- Be careful about the depth of the materials that you use. (If they are excessively high or deep, they may rip the sheet or make holes caused by tension lines.) To help prevent this, you can use felt or foam rubber between the press or roller die and the relief materials.

▶ Papers with relief textures made using moulds of different materials such as wood, plastic, and polyester resin

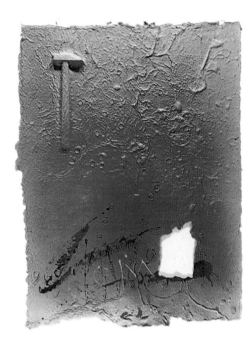

▲ Dai-Bih-In, *Untitled.*
In this piece, the shape in the upper left was created the instant that the sheet was formed—the pulp was completely hydrated. The wet sheet was placed on top of the object to take on its shape.

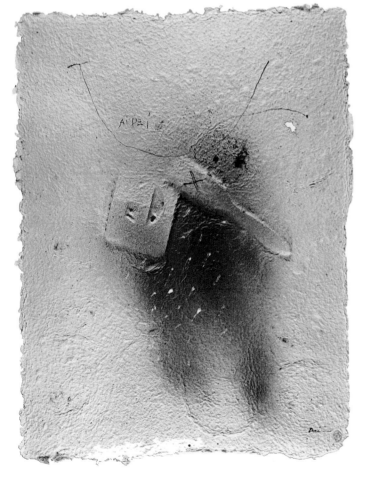

◀ Dai-Bih-In, *Untitled.*
This imprinted piece was made using paper produced by the artist. A native of Taiwan, Dai-Bih-In is a master of papermaking. He uses Oriental concepts of energy and matter in the creation of his works in which paper is used as a medium for aesthetic expression.

Air Drying

If a paper is allowed to dry without pressing, the finish will be as rough as possible for that particular paper. The fibers contract when they dry, and they form fairly noticeable lumps, depending on the type of fiber used and the degree to which it has been refined.

This natural phenomenon can be used to create some nice textures. Within limits, you can manipulate the pulp of the paper that has been formed on the felt, as if you're dealing with some other malleable material such as wax or clay. You can tear it, scrub it, displace it, and so forth, using your hands or any other instruments such as wires, pieces of wood, and ropes.

Many artists have used this technique to achieve very expressive results. A general term for this type of work is *paper nu* (naked paper). This term, coined by Narcis Banchs (Gelida, Spain), seeks to define the material as personal and subject to the artist's influence.

MAKING paper by hand hasn't changed that much from when it began. The same methods, equipment, and tools are still used, with a few modifications brought on by technological advances. The main variations include the use of plastic, with all its advantages as a water-resistant material; new, highly efficient glue and coloring agents designed exclusively for the paper industry; and finally, the Hollander beater. This machine, which was invented in the sixteenth century, makes it possible to prepare the fibers by precisely controlling the level of refinement.

Making paper by hand doesn't require a great investment of money on the part of anyone who wants to do it. As you'll see in this chapter, it's possible to set up a small workshop even with very limited equipment, much of which is available at home. A kitchen blender, a tub of some sort, a fan, and a few boards and screws are all you need to get started. Later, if you want to expand and improve your resources, you can make or buy a press, find a Hollander beater, and set up an area for drainage—in essence, set up a small paper mill similar to traditional mills.

Materials
and Tools

A Traditional Paper Mill

A common denominator of all old paper mills was their proximity to water. The water served several functions: in conjunction with waterwheels it provided a source of energy for driving the refining and smoothing hammers; it was used to wash, refine, and bleach the pulp; and it was an indispensable medium in the process of forming the paper into sheets. As a result, in geographic areas close to water, you can still see buildings that served as paper mills. One common, recognizable characteristic of a paper mill is a large number of windows in the upper stories where the paper was dried.

In the past, a paper mill was not only the livelihood of its workers, but their entire life. People often worked, ate, and slept in the mill. Entire families, including the children, were involved in producing paper at a dizzying pace; and, frequently their living conditions were unhealthy because of the humidity, the lack of light, and the deafening noise produced by the hammers beating inside the vats.

In a traditional paper mill, the paper was made in the lower level or basement where it was possible to activate machines that worked from the force of water falling from a higher level. This water came from a sluice: a natural or artificial channel. Because the work was done in the basement, there was no natural light, so oil or electric lights were used, depending on the time period.

In addition to the area set aside for the machinery, this lower level contained three clearly delineated work areas: one for preparing the raw materials for refining (usually rags); the area where the sheets of paper were formed; and finally, the gluing area.

The first and second floors were almost always reserved for the owner's living quarters; and, in many mills, the workers also lived there. These floors contained the kitchen, a common dining room, bedrooms, and a large space where the dried paper was handled. This area was used to select and package the paper for its ultimate distribution.

The upper story was reserved for drying the sheets. For this reason, there were many windows on this floor that were used for controlling the air circulation. Some mills had more than one floor for drying the paper, and some had a garden area that provided food for the people who lived and worked there. On the following four pages, we'll use the Moli Paperer de Capellades in Cataluña, Spain, as a model for understanding the structure of a traditional mill and its appearance.

The industrialization process in large-scale papermaking that happened in the nineteenth and twentieth centuries was difficult for the traditional mills. The old buildings in which they were set up were too small for the large machines, and many mills had to give up production. In conjunction with this development, many workers moved to large industries that were better prepared to supply the huge worldwide demand for paper.

▼ The Moli Paperer de Capellades Museum in Cataluña, Spain. This eighteenth-century mill, in which paper is still made by hand, is also a magnificent museum. Its state of preservation places it among the most important European museums devoted to papermaking.

▶ Layout and organization of a traditional mill. The number of stories in the building and the equipment varied according to production capacity.

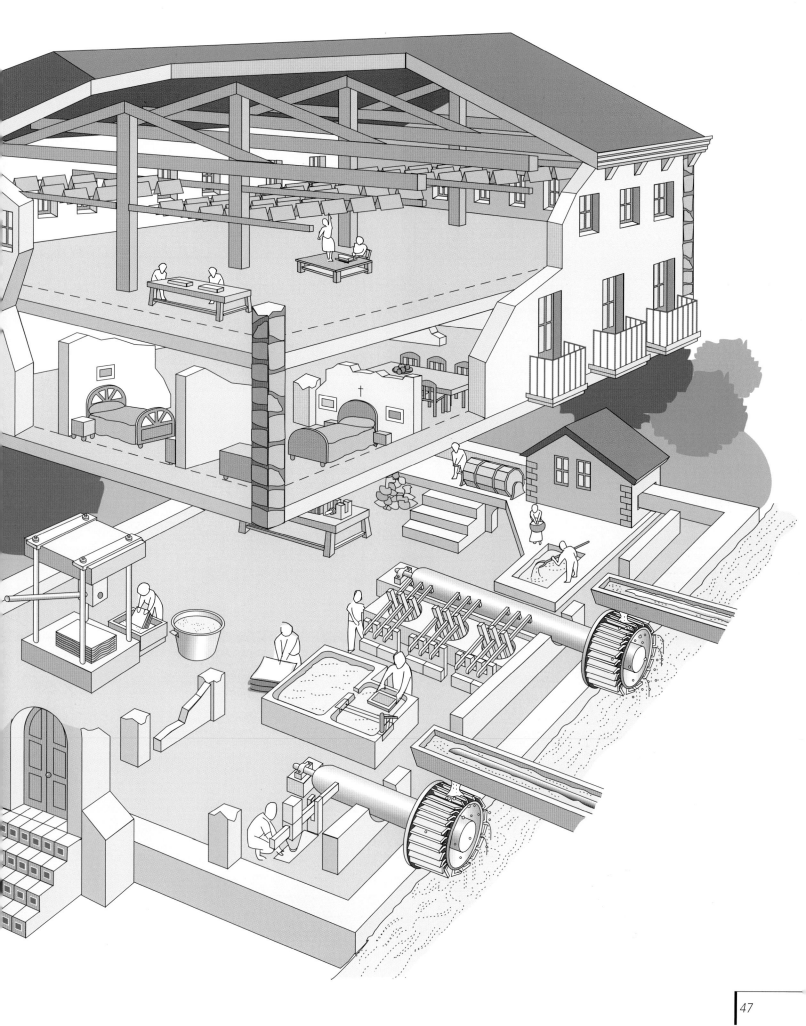

◀ This is the area of the mill where the raw materials (usually rags) are mixed with water, or water and calcium carbonate, to ferment before shredding. The fermentation removes the oil and softens the fabric, making it easier to process.

◀ This is the area where the rags are prepared. First they are winnowed by tossing them in the air; then sorted by category on the cutting table; then torn apart. The buttons and stitching are removed at this point. Then the rags are agitated in a fulling machine. Other raw materials, such as ropes and sandals, are cut up on a wooden block using an axe.

▲ This area is devoted to preparing the paper pulp. There are three different types of hammers for shredding, processing, or refining materials that is transferred to different vats as the materials is reduced and separated through hammering. The bottom of each vat has a metal plate in it for absorbing the blows of the hammer.

◀ The source of the energy is water released by raising the gates to the channel (sluice) and allowing it to flow from the sluice to the waterwheel, providing the power for the hammers.

THE MILL WORKERS

Several types of workers were needed in a paper mill. Depending on the size of the mill, the workers either specialized in a single phase of paper production or performed several functions:

■ The foreman was responsible for the work of the entire mill. In some cases, he was a non-salaried employee who worked on commission based on sales of bales of paper. Sometimes he was also responsible for paying the workers and other expenses.

■ The "vat man," the worker with the greatest skill, was responsible for making the sheet of paper in the vat using a mould.

■ The "coucher" took the mould from the vat man, and transferred the sheet of paper onto a felt placed on a bench. Then he returned the mould to the vat man, and put another felt on top of the sheet.

■ Another worker (the "lay man") was charged with separating the sheets from the felts when they were removed from the press. Then the sheets were placed on the laying bench.

■ The vat attendant took care of the vats, cleaned them, filled them, and mixed in the calcium carbonate.

■ The "hanger" was the worker who hung up the paper in the upper story of the mill. This job was almost always done by women. Children carried the paper from the basement to the loft and delivered it to the hangers.

■ Apprentices who served various functions might wait for two or three years for an opening.

■ Month-to-month contracted workers ate and slept in the factory or mill.

■ Temporary workers traveled from mill to mill to hang out paper in the loft when extra help was needed. These workers were paid by the ream.

■ Other specialized workers were used to apply sizing, trim edges, smooth, sort rags, and so forth. The jobs that these workers did depended on the volume of the work and the number of employees.

▲ The vat is the nerve center of paper production in a mill. This is where the vat man and the coucher work side by side. A fork is used to mix the pulp in the vat, and the coucher returns the mould to the vat man through a bridge or drainer.

▼ The press is indispensable in a mill. Next to the press is the laying bench, where the sheets of paper are placed as they are separated from the felts.

◄ The paper is hung up in the loft of the mill to dry naturally, and the air currents are adjusted by opening or closing the windows. The bench used by the hangers is on a cart to make it easier to move around.

The kettle is used to prepare the sizing using remnants and waste from tanners and butchers. The paper is soaked to moisten it, then a lead weight is put on top of it to keep it from floating and help it absorb more sizing. Then it is pressed to encourage dispersion. A basin is placed below the press to catch any surplus.

▼ The smoothing mallet is powered the same way as the vat hammers, and the papers are positioned and moved on a yoke or metal plate to receive consistent strikes.

▼ On this trimming bench, a knife such as the one shown is used to cut the edges, along with any pages that stuck out of the bundle of one or two reams. The wooden covers, or wings, on each side of the bench, are opened out to collect the trimmings. Once the edges are trimmed, the edges of the paper in the bundle are worked over with a pumice stone, a file, or a ball of paper to smooth them.

Setting Up a Small Workshop for Making Paper

You don't need a large space or costly equipment to make paper; but, keep in mind that the quality of the paper and the amount that you can produce depend on your equipment. The main requirement for setting up a paper shop is space. Today, you can use portable studio equipment, but it's preferable to work with basic equipment in a designated space.

Space Requirements

There are two basic requirements for a papermaking studio: adequate ventilation and a good drainage system. Because water is the principal medium used in papermaking, you'll need good ventilation, preferably natural, to avoid a buildup of humidity. For good drainage, it's best to have a setup such as a huge washtub with a drain beneath it in the floor. The drain allows you to work with water without complicated plumbing, since the excess liquid goes directly into it. It's also easy to clean up using a pressurized hose. The optimum setup is having the drainage area close to a window. For this purpose, a garage or a partially enclosed outdoor patio are good possibilities.

Take a look at the ideal paper workshop set up in the Barcelona School of Arts and Crafts, pictured below. During classes, this space can accommodate twelve students divided into three groups of four. There are three work areas located inside the water area (vat and couching bench). The space is more than adequate for one person.

Away from the area where water is used, there should be an area designated for work with dry materials. It can contain a forced-air dryer, if needed, and a storage area for materials such as sizing, dyes, and fiber. It might also have shelves for storing the dry paper.

▼ Parts of a workshop: Drying rack (1), felts (2), moulds (3), water source (4), water area with drain (5), press (6), ventilation (7), containers (8), vat (9), flat surfaces (10).

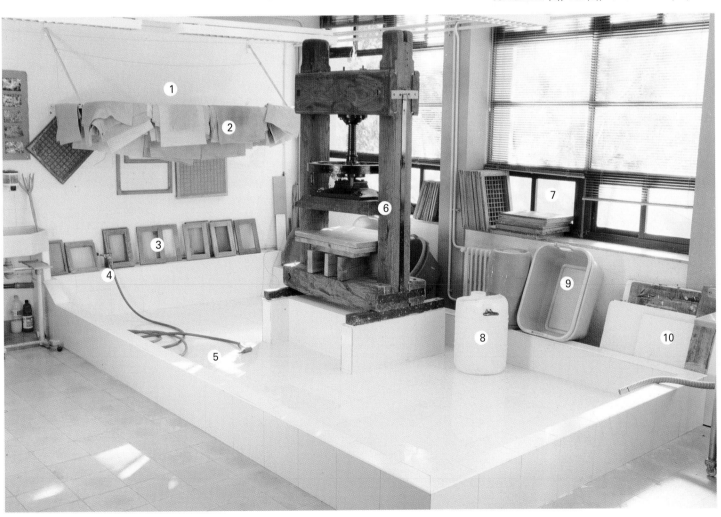

Refining Instruments

The first pulp-refining systems used by the Chinese were hand operated. Animals were used to assist if auxiliary power was needed. The Arabs were the first to employ hydraulic energy, using water to activate the beaters. A wooden cylinder around 14 inches (35 cm) in diameter, known as the lifter, was turned by the water wheel. As it turned, it raised the refining mallets by means of installed cams. The mallets—thick, rectangular blocks of wood—would then be allowed to fall onto the rags in the trough, striking them with the tremendous force of their weight. The bottoms of the mallets were either spiked or smooth, depending on their function. The bottoms of the first mallets, which were used on the rags, had spikes with sharp, tapered edges. The second set of mallets for working the pulp had blunt or flat-tipped spikes; and the third set, for refining the pulp, were smooth.

The rags were chopped up or the pulp was refined in troughs or sinks. In the old days, there were three oval troughs made of stone (mortars) in which the three steps described above were carried out. The first was the rag trough, where the shredding was begun. In the second, the grinding process continued, and the fibers were separated and washed. The third trough was for refining and producing pulp ready for making paper.

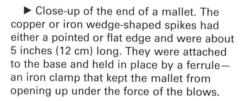

▲ The major drawback to this old refining system was the time required for processing; it could go on for as long as three days. And, imagine the horrible din that all the hammering produced. Beaters pounded the rag with great force against an oval-shaped piece of metal on the bottom of each trough.

▶ Close-up of the end of a mallet. The copper or iron wedge-shaped spikes had either a pointed or flat edge and were about 5 inches (12 cm) long. They were attached to the base and held in place by a ferrule—an iron clamp that kept the mallet from opening up under the force of the blows.

▲ Ancient mortar used for beating and grinding fibers by hand

◀ This little gem of industrial archeology is a handmade Hollander beater by the craftsman Lluis Morera.

This system, introduced by the Arabs, was used until the end of the seventeenth century. With the advent of the Hollander beater, the work that previously went on for two days full of ceaseless noise, could be accomplished in less than two hours.

The Hollander beater, also known as a cylinder mill, shredder, or refining cylinder, is a machine that accomplishes the same shredding task as mallets. With the beater, one is also able to control the force applied to the fibers.

A Hollander beater is made up of the following components: an oval reservoir with a central partition, a beater, a plate on the bed of the beater, and a top. The design of today's Hollander beaters is essentially the same as that of the original ones, since the same process is still used. The fibers are circulated through the channel of the reservoir, and as they pass through the beater they are softened, beaten, or blended, depending on the pressure applied to the plate on the bed. This pressure is controlled by screws that raise or lower the beater over the plate. The machine is usually powered by an electric motor, but there are some manual beaters. This system heats up the pulp a bit, but it's important to keep the temperature below 113°F (45°C) to avoid damaging the fibers.

There are also some domestic appliances that can be used to refine fibers. These include kitchen blenders and paint mixers used with an electric drill. Although useful, they're not the best, since they can cut the fibers at the same time that they shred them. If you're using one of these options, you should file down the blades a bit to reduce their cutting ability so that the beater can separate the fibers without cutting them to pieces.

► Here's a steel Hollander beater made by Soteras. It holds a little less than 2 pounds (800 grams) of dry pulp, or about the right amount for the needs of a small studio.

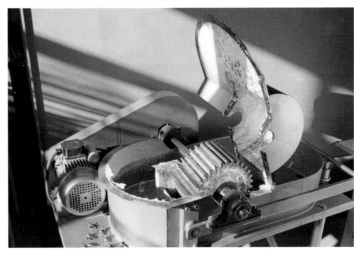

▼ The bed is located under the beater and is attached to the channel. It's a geared piece of iron or steel with flutes or blades that mesh with those of the beater. When the fibers pass through the small space created between the two components, they are shredded and thinned without being shortened.

► The cover helps to avoid splashing. Lifting it up exposes the beater, which is a wooden, iron, or stone cylinder in the Hollander beater fitted with bronze or steel blades arranged parallel to the axis on which it rotates.

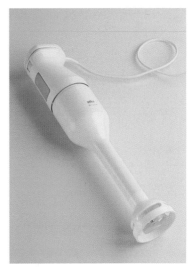

◄ This conical refiner is another system similar to the Hollander beater that is used more frequently in an industrial setting. In this instance, the beater has a conical shape. This cone also performs the function of a pulper to break up the pressed, dried pulp.

► If you use a hand-held blender, you have to exercise care in two critical areas: avoid burning up the motor by keeping it in operation only for short periods of time, and never submerge it completely. This is an electrical appliance that you're using with water, so be cautious!

Presses

Even though the fibers in a sheet of paper interlace naturally during the drying process, pressing increases their cohesion. The amount of water is reduced; and, consequently, air and space between the fibers is eliminated. Pressing also accelerates the dehydration of the fibers; and, therefore, the drying time is much shorter than if the paper were spread out on a couching pad after it is taken out of the mould.

Presses have been used since the invention of paper, even though the methods have evolved. For example, in many settlements in the Far East, pressing is done with weights—almost always rocks—on top of the pile of damp felts. Throughout the world, it's most common to use a press, whether of the turn screw or hydraulic type.

A press is made up of two basic components: a solid frame, where the materials to be pressed are placed, and a movable part that applies the pressure. The frame for pressing paper must withstand strong pressure between its crossbeams and can be made of wood, iron, or a combination of the two materials. If it's made of wood, it should be a hardwood of adequate thickness. The frame opening determines the paper's size. It's useful and advisable to have an opening of about 2 feet (60 cm), since that will make it possible to make larger sheets of paper.

▼ Wooden press from the twentieth century. This press works by means of a wheel that turns the screw. Since the screw doesn't reach the bottom of the press, planks and boards have been used to increase the thickness of the pile so that it fills the space.

▲ Eighteenth-century sizing press belonging to the Moli Paperer de Capellades (Cataluña, Spain). This type of mechanical press is also known as a screw press because of the way it's constructed. The head of the turn screw has holes in the sides for inserting the turning lever. The longer the lever, the less effort is required to apply pressure.

▼ A small iron press is perfectly suited for pressing. After using it, it should be dried to prevent rusting.

The movable element used to apply the pressure varies according to the type of press. In the case of a mechanical press, the force is produced by a thick central screw turned by a lever or a wheel. On a hydraulic press, the pressure is supplied by a hydraulic piston inside a cylinder filled with oil or water. The piston is activated electrically or mechanically and extends in response to pressure changes in the cylinder. An ideal pressure for making paper falls between a total of 10 and 20 tons, but you can also get by with less.

Pressure can be defined as the coefficient between the force applied vertically to a surface and the area of the surface. This means that a press that can exert a total pressure of five tons will press a pile of small format papers more effectively than it will a pile of large format papers, because, in the latter case, the pressure will have to be spread out over a larger surface.

With a mechanical press, all the force is concentrated at the point of union between the nut and central screw. The greater the number of threads in contact, the greater the resistance. For that reason, very powerful mechanical presses have screws of large diameters with a low thread pitch so that there are more threads in contact with the nut. The screw is also usually very large and is integrated into the upper cross member of the structure.

◄ A manufactured, contemporary cast-iron press with a system that increases the pressure considerably and easily when the wheel won't turn any farther

▲ The pressure exerted by this hydraulic press removes all the water from the paper so that it's slightly damp when removed from the press. Hydraulic presses create the greatest pressure.

► This nice homemade press costs very little in relation to its effectiveness.

► Here's a press made for use at home that can be built with few resources. Even though the pressure generated is minimal, it removes an adequate amount of water. The planks help spread out the pressure from the screw clamps.

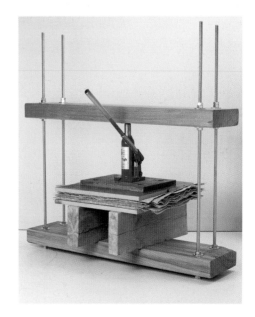

How to Make a Simple Press

You can build a fairly simple press that's effective and doesn't require an investment of a lot of time or money. You'll need to build a strong frame and buy a common hydraulic jack made for lifting cars, vans, or trucks from an auto parts or hardware store. These jacks are a great buy for the amount of pressure they generate.

The construction of the frame is such that it can be assembled and disassembled at will to regulate the distance between the cross members. That way, if you're accustomed to working with small piles of paper, or if you press several times during a session because you're working with different sizes and formats, you can lower the upper cross member. If you plan to press a tall stack just once, you can raise it accordingly. In any case, the cross members should always be perfectly parallel and level; otherwise, the press can be damaged, and the sheets won't come out of the press uniformly dry.

► **1.** For this press, you'll need two planks, four ½-inch (14-mm) threaded rods, 24 nuts, and 16 round washers. Begin by drilling four holes of the same size as the steel rods in the planks, keeping them perpendicular, and in precisely the same location in each plank. (You can drill the holes in the first plank, then use a pencil to mark the location of the holes on the second plank.)

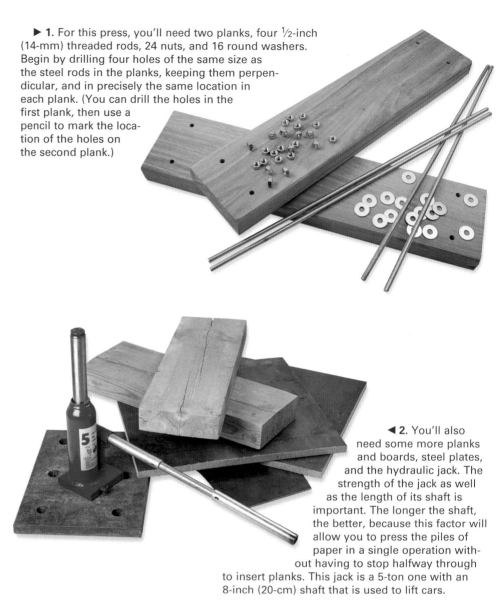

◄ **2.** You'll also need some more planks and boards, steel plates, and the hydraulic jack. The strength of the jack as well as the length of its shaft is important. The longer the shaft, the better, because this factor will allow you to press the piles of paper in a single operation without having to stop halfway through to insert planks. This jack is a 5-ton one with an 8-inch (20-cm) shaft that is used to lift cars.

▼ **3.** Find the location of the inside washers and nuts first, and screw them into place. Once the wood is in place, screw the washers and two outer nuts into place. You'll use two nuts on the outside instead of one, since all the force from the press is exerted at these points.

▼ **4.** Once the two planks are in place, adjust the distances as needed using a tape measure to be sure they are all the same.

The frame of the press is made of wood and steel. The wooden cross members should to be thick and made of strong wood. The best wood choices are strong, dense woods such as ironwood, mahogany, or oak used in the areas of construction where great pressure or friction is generated. Other strong woods can work also, such as cherry and maple, but they're not the first choice for larger cross members since they might not be able to withstand pressure continuously. Use less dense woods for narrower frames that have a smaller opening. (Avoid soft woods such as pine, spruce, birch, black poplar, willow, and hemlock.) The uprights are threaded steel rods that are ⅜ to ½ inch (12 to 14 mm) in diameter.

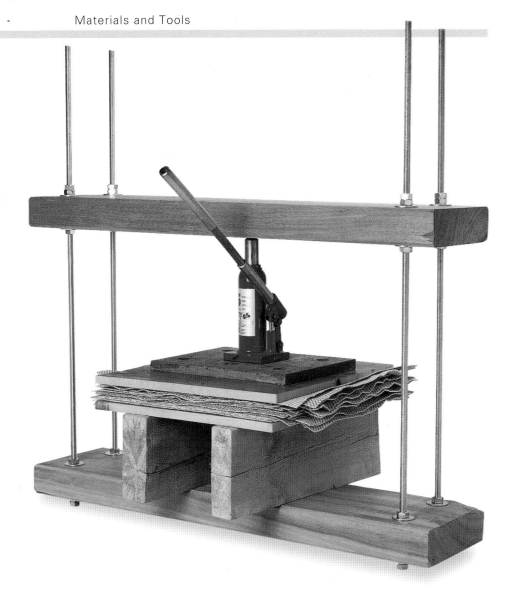

► **5.** Take a look at the finished press with all its components in place. You'll also need a couple of planks and some steel plates to spread out the pressure from the jack, or you might break the planks and damage the upper cross member.

▼ **6.** To apply pressure with this press, you may need to add extra planks and boards to elevate the pile of papers if the distance between the jack that crossbeam is too large. You can also undo the nuts and lower the crossbeam if needed, keeping the beam in the same level position as the crossbeam on the bottom. After this is done, place the hydraulic jack between two steel plates and pump the handle.

▼ **7.** When the jack won't move any more, the stack is under tremendous pressure, and all the water has been squeezed out. At this point, release the jack. Almost all hydraulic jacks release by using the opposite end of the lever on a small key located in the base of the main cylinder. To return the piston to its original position, press down on it lightly with the palm of your hand.

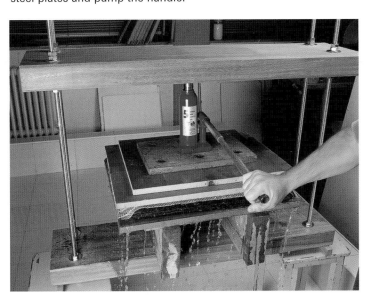

The Drainage Area

Water is the main element used to make paper: It's used in washing and cooking the fibers, beating them; mixing the sizing, glues, and dyes; and as a vehicle for keeping the fibers suspended in the vat. Paper is formed through the use of water, which hydrates the fibers and enhances their cohesion.

The water used for papermaking should be free of impurities and residues, especially organic ones that could end up in the paper and damage it. In the mills, water was channeled and distributed according to various needs. Part of it was used to power the wheels that activated the mills. The rest of it was directed inside to the vats and troughs, filtered by wicker strainers and filters filled with pebbles and sand to purify it.

To avoid rust and corrosion, it's important to be alert to the effect of water on the materials and equipment. Rust and corrosion can be easily transferred to the paper and ruin it. Because of this, you should dry the equipment periodically and make sure that you have good ventilation and dehumidification in your work area.

▲ The water that drains out of the paper in the mould is often a milky color because it contains small, fine fibers. By using very fine mesh in the mould, such as the type used in silkscreen printing, you can avoid losing these fibers and produce a fine, strong paper.

▼ A hose is the perfect accessory for keeping the work area, the moulds, the vats, and other equipment clean and in good condition. A hose takes water right to the area where it is needed, without having to move things around. Heavy vats or troughs filled with water are extremely hard to move.

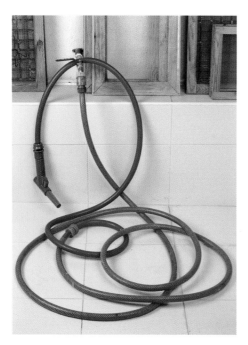

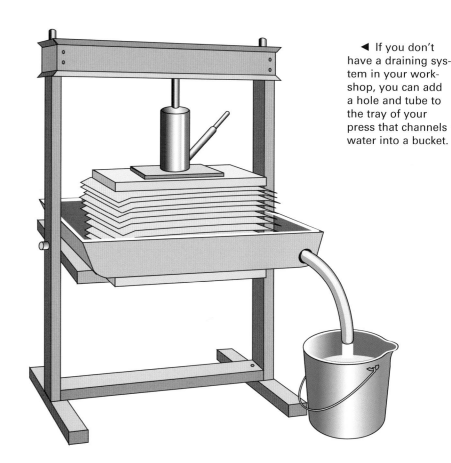

◄ If you don't have a draining system in your workshop, you can add a hole and tube to the tray of your press that channels water into a bucket.

Vats hold the liquid pulp used for making paper. In their original form, they were round or oval, and very similar to tubs used for wine. Vats can be made of wood, stone, construction materials, or plastic in shapes that facilitate water circulation and cleaning.

The size of the vat needed depends on the normal production of the workshop and the space available. A large vat allows making many sheets of paper without having to refill it, since it's possible to maintain the same paper thickness for a long time. The ideal capacity is 25 gallons (100 liters), and the common dimensions used in a small workshop are about 18 x 24 inches (46 x 61 cm) with a minimum depth of 12 inches (60 x 45 x 30 cm). To make it easier to empty, it's a good idea to install a drain in the bottom of a vat that's any larger than these dimensions. You might also want to place the vat on a high, wheeled cart that you can comfortably reach and move around easily.

Historical engravings show that some older vats were heated, because people thought that hot water facilitated formation of the sheets. For that reason, a copper, wood-fired oven was often fitted into the bottom of the vat or a copper tube circulating hot water placed inside the vat.

► In addition to the vats, several types of plastic measuring cups are needed for spreading out the pulp, applying dyes and sizing, and so forth.

▼ You can't make large sheets of paper without a large vat. If the vat is fairly large, it needs to have a drain and wheels. This one, for instance, can be used to make paper measuring 40 x 28 inches (100 x 70 cm).

► The most common types of vats used in small workshops that can be bought at shops specializing in plastic items

◄ You may want to shield yourself from water by wearing rubber boots and a waterproof apron. In warmer weather, wear waterproof sandals.

► Vats used for making paper used to be similar to barrels—an old but effective system.

Drying Systems

An important aspect of papermaking involves drying the paper. Paper that has just been removed from the press still contains 65 to 70% water. The water level needs to be around six percent for it to be considered dry. There are lots of factors to consider about drying. For one thing, paper fibers don't cease being plant fibers after they dry—they still absorb humidity and adapt to environmental conditions like wood does. In other words, the fiber can expand and contract, change shape, or swell up. As a result, these changes need to be controlled as much as possible.

▼ Detail of the rollers in a drying loft

▲ In present-day drying lofts, the sheets are not always hung from wires. Other systems are also used, such as this one, which uses roller-like clamps for holding the paper.

▲ A drying rack such as this household clothes rack can be used for small amounts of paper.

▶ Engraving number 13 from De la Lande's treatise on the art of making paper, 1768

Traditional Methods

The oldest method of drying paper uses the sun's natural heat, and this way is still used with many types of paper in China and Japan. Another traditional drying process—air drying paper in a covered area—involves no pressure, heat, or interference from the elements. Signatures of paper or individual pieces are hung on ropes in the loft or top floor of the paper mill. In this process, the paper dries from the outside in; i.e., the outer layer dries first, and the remaining moisture is concentrated in the interior. Since paper readily absorbs moisture, the loft must be well-ventilated to reduce the air's relative humidity. The drying process in a loft usually takes two to three days.

In the old days, hanging the paper to dry in the mill loft was done by women workers. They moved the paper along on benches, and hung up the sheets in signatures using a T-shaped hanger. This delicate operation had to be carried out with great dexterity, since the damp, tender paper was easy to tear or distort.

▲ Hanging signatures of paper on ropes

▲ A T-shaped hanger eliminates the need to handle paper by hand and helps prevent tearing damp paper.

▼ Drying grates, which are used in graphic industries as well as engraving shops, make it possible to air dry many sheets in a limited space.

▲ Paper air drying in Bhutan. This system is similar to a traditional one used in Nepal. The heat of the sun speeds up drying and whitens the paper.

Current Methods

Today, industrial paper mills use efficient drying methods that suit their commercial needs. The paper is often passed over a felt cylinder that's steam-heated like an industrial iron. In smaller, craft-oriented mills, the paper is dried under pressure before forced ventilation is used, producing excellent results.

One simple method of forced drying uses corrugated cardboard and a fan. To do this, layer pieces of damp paper between sheets of cardboard and weigh them down on top before a fan is aimed at the stack from one side, forcing air through the corrugations in the cardboard, and gradually dehumidifying the pile. Cover the pile with a sheet of plastic, leaving the intake area open for the fan and the front of the pile on the other end open for circulation. The process can take a couple of days, depending on the size and thickness of the sheets. To avoid burning up the fan's motor, turn it off and give it a break after a few hours.

This method doesn't take up much space, the sheets come out flat and smooth, and there is none of the shrinkage associated with air drying, since the weight keeps the sheets from contracting. The only disadvantage is that the weight considerably reduces the texture of the paper, especially if the texture is very noticeable or the paper is crimped.

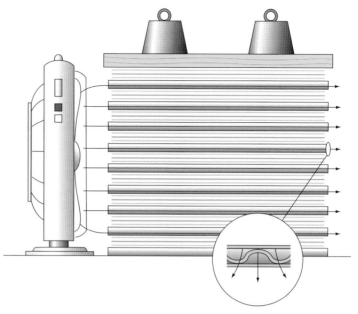

◄ To air-dry paper between cardboard sheets, the pieces of cardboard should all be the same size, piled up, and oriented in a direction that allows the air to circulate through them. Cover them with plastic, leaving the ends open so that air escapes through the corrugated openings. Keep the fan's air intake clear so that the motor doesn't burn out.

► **1.** After making a pile of papers and cardboard, cover it well with a large plastic sheet. Leave the front of the stack open as well as the air intake for the fan, so that it can take in air and force it through the corrugations in the cardboard.

▼ **2.** Adding weight to the pile stabilizes it while acting as a press during drying, and it also keeps the paper from wrinkling and shrinking.

▼ **3.** When the process is finished in a few days, uncover the pile and remove the dry, smooth paper. If the paper is still a bit damp, cover it up again and increase the drying time slightly.

A mould is used to form a sheet of paper. It's composed of a wooden frame with a mesh sieve and a cover, or deckle. In the past, craftsmen made moulds as a trade, but that profession is now disappearing. However, carpenters in areas with a tradition of papermaking still make moulds that resemble the older ones.

▲ When a mould is turned over you can clearly its construction—this one is made of wooden strips to which the mesh is secured to keep it from loosening and distorting during use. The strips help to even out the sheet's thickness during formation. Without these, dips in the mesh would cause the sheets to be thicker in the center, leading to wrinkling at the edges during the drying process.

The Main Section of the Mould

The main part of the mould holds the pulp on the screen while the water drains off. The first Chinese screens were made of horsehair, silk, or bamboo. Later, the Arabs used bronze wires for this purpose. By the mid-eighteenth century, wire mesh was used. The first meshes were made of woven copper wire; but today they are also made of steel, galvanized iron, and synthetics that work extremely well.

The structure of a mould is very straightforward: The wire mesh is secured to the frame with nails and is almost always reinforced from underneath with strips fitted inside it. A small strip of copper or brass plate running around the inside of the frame covers the nails, holding the mesh and its rough edges in place to prevent them from raveling. When the form has a watermark, it is threaded to the mesh with tiny copper wires. Naturally, all the materials should be water-resistant. The wood can be varnished or waxed to seal it, and the metal should be rustproof.

Laid moulds have a special weave made from tiny rods interlaced with fine wires. The rods, commonly made of copper or bronze, are called stitching, and the wires that hold them together with a simple slipstitch are called reglets. The laid screen is sewn to the strips or wooden cross members. In Oriental laid paper forms, the reglets are customarily made of bamboo.

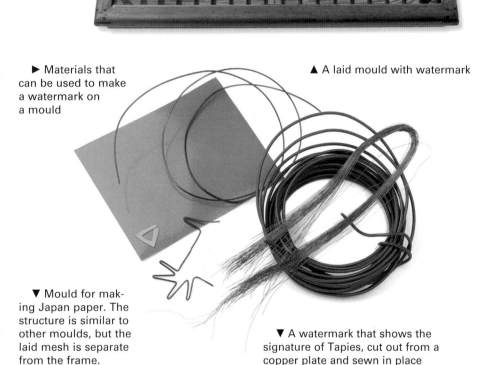

► Materials that can be used to make a watermark on a mould

▲ A laid mould with watermark

▼ Mould for making Japan paper. The structure is similar to other moulds, but the laid mesh is separate from the frame.

▼ A watermark that shows the signature of Tapies, cut out from a copper plate and sewn in place

The Deckle

The wooden deckle covers the mould's frame and screen to hold the liquid and control drainage while the sheet is being formed. It almost always has a groove to make it fit the form more precisely, but this feature isn't essential. The depth of the frame controls how much water it can hold. For instance, a very high frame allows one to make thicker papers than a low one because it holds more water. Because the draining time is prolonged, there is time to integrate added materials (such as flowers and aquatic plants) in the paper.

As we saw in the preceding chapter, the deckle is crucial in the formation of frayed or deckle edges. Rough edges and defects in the fraying are often due to improperly fitted deckles. Therefore, it's preferable that a deckle has an alignment groove to prevent movement inside the vat. The deckle is also indispensable in making special-format papers, such as envelopes, since it defines the actual shape of the sheet.

▲ To make small cards or sheets of paper, a large form is used and thick adhesive tape (such as insulating tape for windows) is added to mark off the dimensions of the papers. This method allows a number of small cards to be produced in a single operation. (It's inefficient to make them one at a time using a small form.)

▶ A frame for envelopes, cut from hard plastic with a coping saw. Hard plastics last for a long time. The thickness of the plastic influences the paper's thickness.

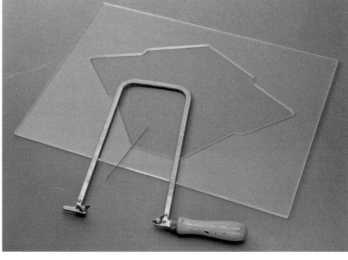

▼ The reverse side of the mould and deckle

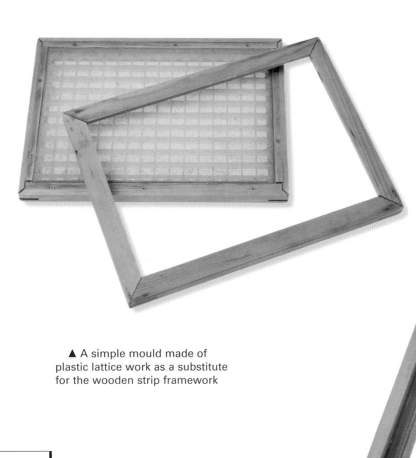

▲ A simple mould made of plastic lattice work as a substitute for the wooden strip framework

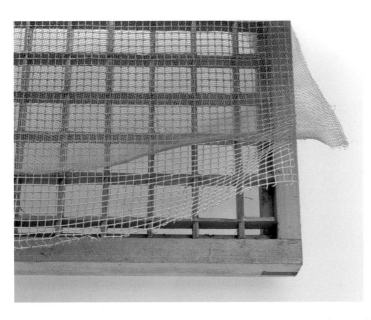

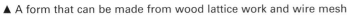

▲ A form that can be made from wood lattice work and wire mesh

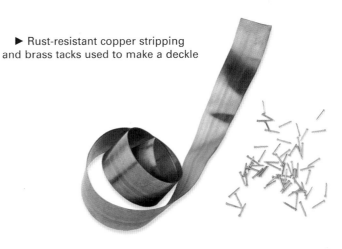

► Rust-resistant copper stripping and brass tacks used to make a deckle

▼ Using a highly refined pulp in a mould with an extremely fine mesh results in very fine papers.

▼ A deckle for making envelopes two at a time

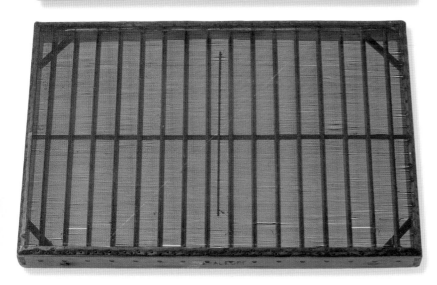

▲ A homemade mould with galvanized metal mesh stapled in place. This is the most elementary form that you can make, and it works perfectly for small format work.

Pulps, Sizing, and Coloring

Beginning with the Arabs, rags were used to make paper pulp. The plant fibers found in the rags were recycled by means of grinding, shredding, and refining. This proved to be a costly production method that was slow and dirty. Also, as the demand for paper grew, there wasn't enough material available to satisfy it. By the mid-nineteenth century, extensive experiments were being conducted with plant fibers to find another route to making paper.

Just as bricks in a walls are reinforced by other materials such as cement, plaster, or steel, the cellulose in plants is supported by other substances. In order to make paper pulp, the fibers in plant tissue must be isolated and separated so that their desired properties fit their eventual use. Pulps are classified according to the type of energy used to break down the connections among the fibers, whether it is thermal, mechanical, or chemical energy—or a combination of all of three.

Today, some factories produce pulp as their main product. Other factories handle the complete papermaking cycle from tree to manufactured paper that is ready to be sold, handled, and printed in book form.

Small-scale artisans who make paper can buy pulp from a factory or small paper mill, and in many places, small quantities of pulp are sold in art supply stores. The other option is to recycle paper or produce the pulp directly from plants or rags.

Preparing Pulp

Paper pulp, whatever its origin, passes through three major phases before being made into paper:

1. Dry preparation of raw materials: Cleaning, dust removal, shredding, and selection.

2. Wet preparation: For mechanical pulp, this stage consists of extracting and classifying fibers as well as optional bleaching. For rags and chemical pulps, this step consists of decomposition through grinding or lye solutions and washing, with optional fiber extraction and bleaching.

3. Refinement: More or less advanced refinement is made depending on the type of paper desired, with the optional addition of sizing and color.

Chemical Pulps

Hugh Burgess and Charles Watt invented the chemical method for separating plant fibers by boiling fibrous stalks in caustic lye to produce what's known as soda pulp. The chemical treatments to which pulp is subjected vary greatly—the most common ones involve the use of soda, bisulfite, and sulfate (which produces kraft pulp).

Chemical pulps are beaten and left unrefined to prepare them for distribution. Sometimes they are prepared dry and sometimes wet, in order to facilitate fiber separation in the pulpers (disintegrators) of paper factories. Small-scale papermakers can buy these pulps in dried form and rehydrate them to restore their properties and smooth appearance. The degree to which they are refined is determined by the factory or mill, and various pulp mixes are made to produce the desired type of paper.

Traditionally, chemical pulps are bleached with chlorine and its derivatives. Because of the pollution caused by chlorine, other bleaching agents are used today, such as ozone, oxygen, and oxygenated water. Because of its purity, chemical pulp is considered to be the highest grade of pulp.

PINE

LINEN

COTTON

EUCALYPTUS

MANILA

Mechanical Pulp

In chemical pulps, nearly 40% of the plant is extracted, but mechanical pulp uses almost all of it. There are two main reasons for using mechanical pulp: its low cost (when energy costs work in its favor) and its responsiveness to printing. As a result, most mechanical pulps are used to make paper destined for the printing press, such as newspapers and magazines. These papers aren't very strong, and they yellow with the passage of time. In order to produce a stronger sheet, pulp containing long fibers from conifers is added to the pulp.

The materials for making pulp, which generally come from wood, are boiled and swell as they absorb water. Then they are beaten, cleaned, and refined.

Coloring Pigments

To eliminate the natural color of the fibers to produce white, it's common to add chlorinated solutions to the pulp. Formerly this involved lime; later on, ash lye was used; and today, various chemical processes are used. In some mills, this operation is carried out in the bleaching room. Once the pulp has been bleached, color can be added. The color can come from a number of sources:

- Natural colorings (tea, coffee, boiled onion skin, etc.)
- Aniline dyes and inks (from food, hides, wood, etc.)
- Clothing dyes (added to the pulp before making the paper)
- Powdered pigments
- Water soluble paints
- Colorings made for paper

Among all of these coloring agents, the preferred ones are the pigments, both natural and designed for use with paper, since they are harmless and clean to use. The color is applied cold, directly in the vat when the pulp is being refined, improving its penetration.

Semi-mechanical and Semi-chemical Pulps

These pulps are derived from parts of both processes—both chemical and mechanical. During manufacturing, the raw plant materials are subjected to a mild chemical treatment to soften them; and later, they undergo a mechanical procedure. These pulps are made with small-diameter wood and quickly-growing plant species such as eucalyptus. Since World War II, the industrial development of such species has played an important role in the paper industry. Since mechanical pulp is of low quality, it is mixed with chemical pulp to produce the required characteristics of strength and the ability to take print.

Sizing

The first types of sizing for paper developed by the Chinese were made from aquatic plants and plant juices. Later, animal starches and gelatins were applied as sizing. Today, artisans use internal sizing that is clearer, longer lasting, and easier to use.

Animal sizing was made by boiling the trimmings from tanneries and the waste from butcher shops in a kettle to reduce them to an oily gelatin. A wooden or copper strainer containing meat trimmings was placed in the bottom of the kettle to keep the sizing from sticking and burning. Sometimes, pieces of rope were used for this purpose.

Next, the paper was submerged and a lead weight placed on top to keep it from floating and help it absorb the sizing. Then it was pressed to encourage even distribution. A basin was placed at the feet of the sizing press to collect the excess.

In his treatise on the art of making paper, M. de la Lande advises against using sizing produced from pigs. He recommends sizing from goats, lambs, and sheep because they produced a whiter sizing. He considered fish to be the best source of sizing. In general, animal sizing lends a yellow cast to the paper, as well as a characteristic shine that's harder and more metallic looking than other types of sizing.

Starch from flour (wheat or rice) or resin are all sizing from plants. Rosin, a type of resin that's extracted from the trunks of pine trees, is refined to make sizing. Used as early as the fourteenth century, the refining process involves separating the turpentine to produce an

◄ Various types of sizing that can be used for paper (left to right): rabbit sizing, the present equivalent of animal glue that's sold in sheet form and is characteristically brown; cellulose glue, which is transparent and diluted; paper sizing (alkyl ketene), which is sold in a dispenser; and fish sizing, which is sold in transparent sheets.

amorphous mass that dissolves in hot water and mixes with paper pulp and certain quantities of caustic soda or sodium carbonate. Today, rosin mixed with aluminum sulfate is used as surface sizing for industrial paper; however, this combination is very acidic and detracts from the permanence of the paper. For this reason, there is a growing tendency to use neutral or alkaline sizing processes, such as synthetic sizing agents.

An emulsion of wax and casein is used to size card stock. Wax is also mixed with starches for surface sizing in order to reduce the size of the paper's pores, and, as a result, the speed at which fluids penetrate.

There are types of sizing that are sold commercially, such as latex, hide glue, cellulose glue, and bookbinding glue. These have to be diluted with water because of their adhesive power. Before deciding to use one of these, it's a good idea to experiment with the proportions to find the best mix. Every type of glue and every brand responds differently.

Many kinds of sizing are spoiled by heat, so it should be avoided. If necessary, it can be refrigerated. Sizing should also be tightly sealed after use.

A different degree of sizing is required for various paper uses (writing, folding, engraving, water colors, etc.). In general, a properly sized paper can absorb water without losing much strength, although some papers require a stronger dose to be sufficiently waterproof.

Other Additives

Other materials are added to manufactured papers for various purposes. The main ones include:

- Sizing agents such as kaolin, calcium carbonate, talc, and titanium dioxide

- Strengthening agents designed to fortify the paper and improve the connections between the fibers. Some are applied dry (starches, gums, and synthetic polymers such as polyamides and latex); others enhance the strength of the paper when it's wet (urea-formaldehyde resins, melamine-formaldehydes, and polyamides).

- Special additives are used in certain papers to produce qualities such as absorbency, softness, dimensional stability, increased rigidity, or fire resistance.

- Control additives applied to the fibrous suspension to facilitate the manufacturing process (biocides, anti-foaming agents, diffusing agents, fiber defloculants, drainage aids, etc.).

◄ Colorings and pigments for paper. Any coloring applied to paper must be water soluble.

Felts are used to support paper that has just come from the mould. They protect the sheet from tearing, facilitate carrying it to the press, and keep the sheet together while it's being pressed. They are usually woven or fulled from natural wools. Some contain a mixture of nylon or polyester. They should have a smooth, uniform weave with no stitching.

Before using them, dampen the felts by submerging them in a separate vat of water. (However, it's not a good idea to allow them to soak in the vat.) When the work session is over, dry them before the next round of use.

Once felts begin to wear out, they are subject to attacks by microbes from sizing that has been added to the paper. Coloring agents also leave traces in the felts and can stain the paper, so the felts should be washed periodically with detergent and chlorine bleaches. If the felts begin to fray, they are unusable because the rough texture of the fibers will stick to the paper fibers, and they can't be separated from the sheets after pressing.

When you remove paper from the felt, vigorously brush off any vestiges of pulp without destroying or fraying the fabric.

Woolen felt fabric is normally used for this purpose, but old woolen blankets, tightly woven and worn carpeting, and kitchen towels also work well. The dimensions of felts vary according to the size of the paper and the press. The average size is about 24 x 18 inches (60 x 45 cm).

A larger felt or stack cover is usually folded and placed on top of the last sheet or paper in the stack before it is pressed.

▶ Boar's bristle or nylon brushes are used to remove traces of pulp from the felts. Clothes brushes can also be used since they are firm and designed not to harm fabric. Never use metal brushes, since they'll tear the fabric.

▼ Felts must be soaked before using them.

▼ Durable felt fabric and carpeting are ideal materials to use as papermaking felts.

▲ Inexpensive kitchen towels can be used as felts, even though they aren't as durable as felt fabric or carpeting. Sometimes the cheapest ones in the store work as well as any.

Other Useful Tools and Utensils

One of the most interesting processes in handmade papermaking is making paper pulp from plants (grasses, straw, or palm fronds, for instance). A chemical process using heat is needed. This process can be done using a pot, some strainers, a hot plate, and an exhaust fan. Depending on the type of plant, you can use garden, kitchen, pruning shears; a hobby knife; or even a garden chipper to shred the plants.

Usually, the shredded plant material is boiled in a solution of water with caustic soda (sodium hydroxide) that serves to isolate the fiber. Sodium carbonate, which is a lot less harsh, can also be used if soda is too aggressive for the fibers, (for instance, the oriental fibers *kozo*, *gampi*, and *mitsumata*).

To smooth out the paper made from plants, you can use a household iron or press the sheets between metal plates.

▲ Caustic soda (sodium hydroxide) and powdered sodium carbonate are two chemicals that serve to isolate fiber when making pulp.

▲ The pH factor has to be controlled when making pulp. It can be tested using strips such as this one that can be dipped into the boiling solution containing the fibers. The strip turns a color that indicates if the pH is acidic, neutral, or alkaline.

▲ Examples of pots that can be used to cook plants and isolate the fibers

▶ This garden chipper, used to shred branches from pruning trees, can also be used to cut up plants before boiling them to make pulp.

▼ A hot plate is an indispensable tool for preparing chemical pulp at home.

▲ A food processor can be used to shred lots of materials that blend into the paper better if they're reduced to small dimensions (dried flower petals, straw, fibers, etc.).

◀ Strainers get a lot of use in making paper, both for cleaning up the fibers and holding leftover pulp when the work session is over.

▶ An iron can be used to smooth out wrinkles and put a satin finish on the paper once it's dry.

Setting Up a Paper Workshop with Limited Means

With some time and imagination, it's possible to set up a papermaking workshop without a large investment by using common, inexpensive things. (There are also small, portable papermaking workshops designed for school use that are available on the market; but this type of equipment is more expensive and not very durable for repeated use.)

If you're working with limited funds, you can use whatever you already have on hand. Perhaps you'll only need to buy a few things to get started. The following list suggests some economical solutions for various needs:

- Paper pulp: You can use recycled paper of decent quality (avoid newspapers, if possible). Tear the paper you plan to use to get a sense of its fiber content: the more visible and longer the fiber is, the better. You can also experiment with plants and extract the fiber from them yourself.

▼ You don't need a large space in which to make paper. You can use a corner such as this one, and keep adding to your suppliers and space as you discover the possibilities offered by this fascinating craft.

- Sizing: Paper sizing is inexpensive and lasts awhile. If you can't find it, substitute latex or cellulose glue, neither of which are costly.
- Coloring agents: Tea, coffee, onion skin, and inexpensive pigments can be economical solutions to coloring.
- Felts: Simple kitchen towels of any brand, worn wool blankets, old cloth sacks, or other smooth materials can be used as felts.
- Vat: A large basin or a sink can be used for sizing.
- Hollander beater: A kitchen blender is used for this purpose in lots of small studios.
- Press: Use two planks and four carpenter's clamps, or a homemade press with a hydraulic jack (see pages 56 and 57 for more information).
- Drying facility: You can use a clothesline and clothespins, or the drying system using a fan and weights (see page 62).
- Mould and deckle: Simple ones can be made with two frames of the same dimensions with metal mesh (screen) stapled to one. To make a deckle for envelopes, you can cut out a silhouette from a sheet of plastic.

Technical *aspects*

S O FAR, we've looked at the materials and the tools used to make paper. Now, you'll learn how to make paper step by step—from the preparation of the pulp through the final pressing of the paper and the drying. We'll also explain how to prepare some special pulps. First of all, you'll learn how paper is recycled, and how it and other types of paper can be made into very attractive sheets. You'll learn how to make decorative papers, such as paper with flowers, that are popular for making stationery, cards, notebooks, and other items. Finally, you'll see how foreign materials can be incorporated into the paper to create a three-dimensional effect.

Handmade Paper and Industrial Paper

Even though paper is still made with the same methods as it was when it was invented in China over 2,000 years ago, the materials and equipment have changed dramatically throughout history. First of all, the raw materials available have varied through time in different geographic regions. People have experimented with every kind of fiber, whether recycled or taken directly from the plants.

The system of making paper by hand—which we'll explain in detail in the following pages—involves three phases. The first is preparing the pulp; the second is forming the sheet until it comes out of the press; and the third is the finish work, which begins with the drying process and ends when the paper is smooth and ready to use.

Pulp preparation begins with choosing the raw materials. If the pulp is made from plants, they have to be shredded and boiled with a chemical agent to isolate the fibers, then washed and lightened in preparation for refinement.

In the second phase, the work can be efficiently divided between a couple of people. In papermaking operations of the past, this phase was carried out by three workers (the "vat man," the "coucher," and the "lay man"); but two people, one to work with the vat and one to couch the paper, make an ideal team. One person can carry out the entire process, but production is slower.

In the final phase, the paper is dried. If the paper is air dried, a second pressing is needed to smooth out the sheets because they wrinkle when they dry and shrink. If forced drying under pressure is used, the sheets will come out smooth and dry. For a satin finish, the sheets have to be rolled or pressed between metal plates. Texture can also be added by crimping. Today, the artisan puts in sizing as the pulp is prepared or prior to making the sheet, so it isn't considered a finishing phase.

Machinery used in paper mills revolutionized traditional methods of making paper in regard to production capacity.

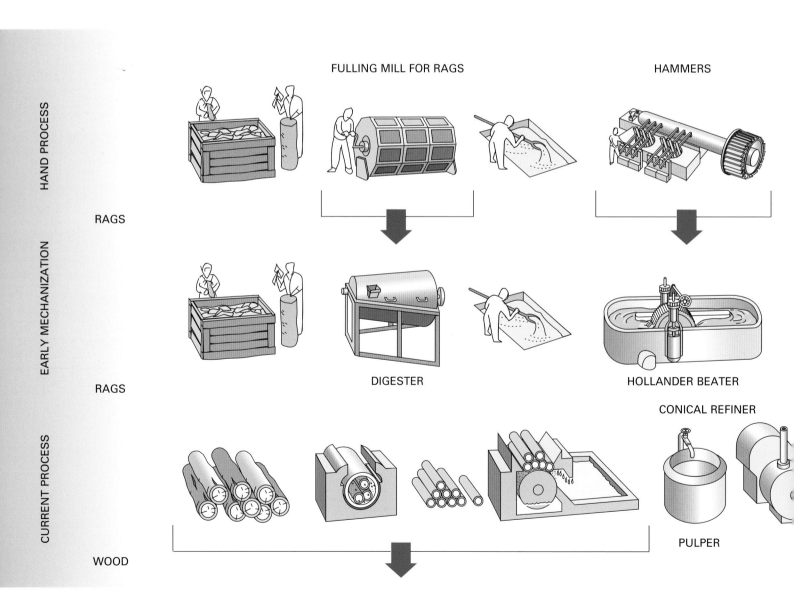

FULLING MILL FOR RAGS   HAMMERS

HAND PROCESS

RAGS

EARLY MECHANIZATION

RAGS   DIGESTER   HOLLANDER BEATER

CONICAL REFINER

CURRENT PROCESS

PULPER

WOOD

THE PULP MANUFACTURING PROCESS

The Hollander beater, discussed in the previous chapter, is one of the most significant additions to the process. The invention of the continuous papermaking machine is even more significant, since it has brought paper production to very high levels.

One of the machines that became quite widespread in many mils was the cylinder or multi-cylinder machine invented by Picardo. This machine did away with the vat man's and coucher's jobs, but not the drying and handling or the pulp preparation. It has a rotating drum that picks up the pulp from the vat and deposits it onto a felt conveyor belt that has rollers to squeeze out the extra water and convey the sheet to the person who lays the sheets. Cloth strips provide a division in the drum for forming several sheets and creating deckle edges; in addition, watermarks can be incorporated in the mesh of the drum. This operation is a partially manual one that facilitates the work without detracting from the handmade quality of the paper.

In 1798, Nicolas-Louis Robert constructed a moving screen belt to make a continuous flow of paper and deliver an unbroken sheet to a pair of squeeze rollers. This idea was perfected in 1807 by Henry and Sealy Fourdrinier and was a major step toward the industrialized manufacture of paper. The machine works by means of a conveyor belt that addresses the various phases of the process along its run. Rollers supply the pressing, drying, and final finish by subjecting the paper to a certain pressure, heat, vapor, and so forth, depending on the stage of the process and the type of paper being made.

This type of machine is used today in industry. Paper factories that use this kind of machine buy the pulp from paper industries and deposit it in pulpers to moisten and stir the pulp before final refining. There are also integrated paper factories whose production process includes cutting the trees, boiling, and bleaching the pulp, refining and preparing it, as well as making papers, finishes, and paper machines.

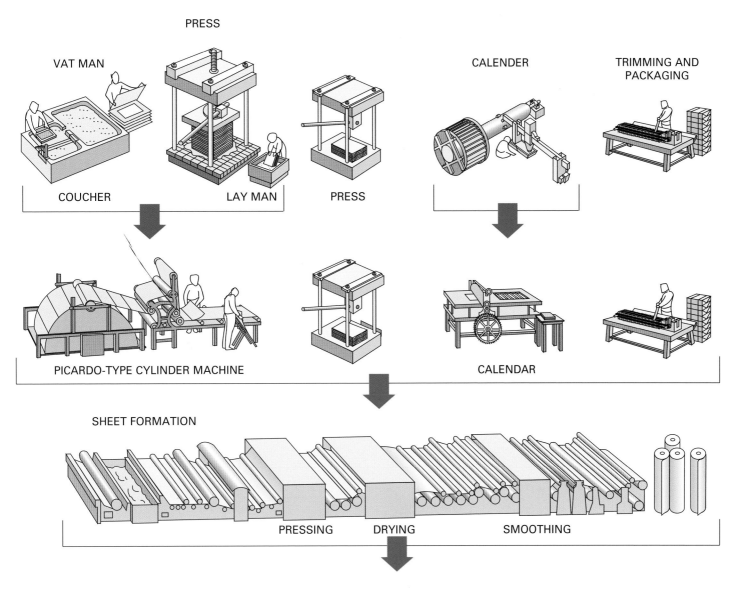

PRESS

VAT MAN

CALENDER

TRIMMING AND PACKAGING

COUCHER LAY MAN PRESS

PICARDO-TYPE CYLINDER MACHINE

CALENDAR

SHEET FORMATION

PRESSING DRYING SMOOTHING

MULTI-CYLINDER MACHINE

Making a Sheet of Paper Step by Step

In this section you'll see how paper is made by hand, step by step, using the descriptions and the photos that correspond to each of the phases of this process.

Preparing the Pulp

The first phase involves preparing the pulp to the point at which it is ideal for refining (depending on the particular type of paper being made). To do this, you must figure out the capacity of the Hollander beater and the strength of its motor, and match the quantity of pulp (or rag) to it. Besides the quantity of pulp, you must also take into account the time required for refining the pulp (or breaking down rags) in the beater. The mixture has to contain enough water, since a very thick pulp complicates and prolongs the refining process. The mixture should be thin enough that the fibers circulate freely between the beater and the plate on the bed (the same principle applies to the blades of a blender). When using a Hollander beater, the beater is adjusted to the plate based on the degree of refinement desired.

If using internal sizing, this is the best time to add it, since it penetrates the fibers much more effectively. If you're preparing pulp for several days, it's best to add sizing at the start of each session so it doesn't lose its effectiveness. If you know your production needs, the pulp can be colored all at once—for example, if you're producing 50 sheets that you want to come out in the same shade of red, the color needs to be added all at once so that you don't have to add more color or adjust it later.

▲ **1.** Weigh the dry pulp prior to refining it. This weight is directly related to the quantity of paper that you'll be able to produce once it is finished and dried.

▲ **2.** Add water to the beater. You don't need to fill it, and the level can be adjusted later.

◀ **3.** Add pieces of pulp to the beater, and allow it to soak for a few minutes to thoroughly soften it.

▼ **4.** Begin by beating a small amount of fiber. Turn on the beater for a short while, keeping the beater high. This step is merely for beating the fibers and separating them, similar to what a pulper does in a factory.

▼ **5.** When the pulp is diluted, lower the beater a bit, and start beating. The pulp, although thick, is free of lumps.

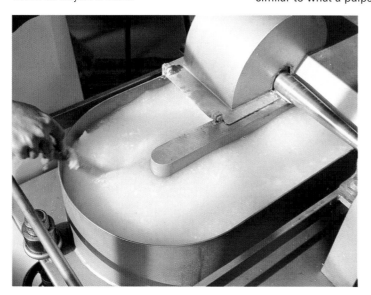

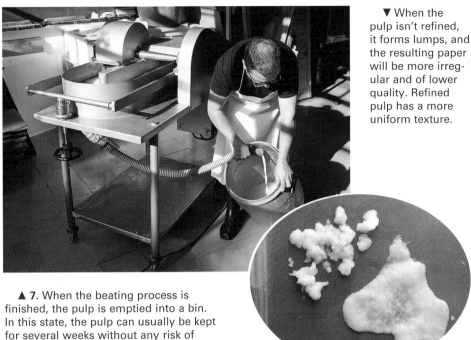

▼ When the pulp isn't refined, it forms lumps, and the resulting paper will be more irregular and of lower quality. Refined pulp has a more uniform texture.

▲ **6.** After a few minutes of beating, you can proceed to the refining stage by lowering the beater more. If you really want to bear down, lower the beater as far as it will go. After a few minutes, the pulp will be soft and refined. The process takes varying amounts of time, depending on the type of fiber and the paper that you want to make (for instance, with cotton linter, it's a good idea not to exceed thirty minutes). Sizing and coloring can be added during this phase.

▲ **7.** When the beating process is finished, the pulp is emptied into a bin. In this state, the pulp can usually be kept for several weeks without any risk of spoilage, if the surrounding temperature isn't too warm. If it's too warm, the pulp could go bad after a week.

Preparing the Vat

Once the pulp has been prepared, you'll need to fill the vat with as much water as needed to make the sheet. The proportion between pulp and water depends on how thick you want the sheet to be. It's common procedure to work a little at a time until the desired thickness is obtained. From this point on, pulp is added to the vat as it's used up. If the vat is small, this will have to be done frequently, and the same is true for making a large sheet of paper.

▲ **1.** Fill the vat with water to a depth of at least 10 inches (25 cm).

▲ **2.** Pour in the amount of pulp needed to make one sheet of the desired thickness.

Soaking the Felts

Before beginning to make paper, you must wet the felts with plenty of water. Any items used to create texture, such as cloth bags or fabrics, must also be dampened. Added materials, such as flower petals, threads, or fibers, should also be moist—with the exception of confetti paper, which can be placed directly into the vat at the start of the session to keep it from softening too much.

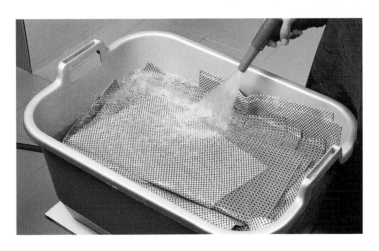

◀ The felts are put in to soak before forming the sheets. Cloths soak up the water quickly, but felt fabric needs a little more time to become thoroughly saturated.

Forming the Sheet

To form a sheet of paper, the pulp in the vat has to be stirred first before the mould is submerged and pushed to the bottom of the vat. Make sure that the frame is properly adjusted before you insert the edge of the mould into the pulp. Move the mould to the bottom of the vat, then bring it up quickly. Shake it gently side-to-side to spread out the pulp and make an even sheet of paper. (A fairly common defect that results is a poorly formed edge resulting from a difference in thickness on one edge of the paper because the pulp wasn't properly distributed.) When all the water has drained off, stop agitating the mould to avoid disturbing the uniformity.

Once the water has drained off, decant any remaining excess water. Next, separate the deckle from the mould, and avoid allowing drops of water to fall off the deckle onto the sheet. If you do, you'll end up with small craters on the sheet. To avoid this, remove the deckle decisively so that it doesn't splash or drip when it moves over the mould.

◄ **1.** If sizing wasn't added during the refining process, now is the time to do it, unless you have decided to use external sizing.

◄ **2.** The fibers tend to sink to the bottom of the vat, so it's important to stir the pulp before making the sheet. Stirring also helps to avoid lumps and uneven thickness from one sheet to another.

▼ **3.** The mould and deckle should fit together perfectly, or the sheets produced will come out distorted or with deformed deckle edges.

▼ **4.** Slide the mould, edge-first, into the vat.

▲ **5.** Push the mould all the way to the bottom to allow movement of the fibers.

▲ **6.** Remove the mould decisively from the vat, allowing the water to drain out through the mesh.

◄ **7.** Move the mould gently side to side, spreading out the fibers uniformly. When there isn't much water left, stop moving the mould to avoid disturbing the structure that's been formed.

▼ **9.** The thickness of the sheet is calculated by eye, and this is how you learn to recognize when the vat needs replenishing. The thickness observed in a moist state is four or five times greater than the thickness of the dry sheet.

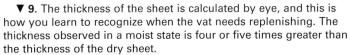

▲ **8.** Lifting up the deckle is a delicate operation, since it often results in drops of water hitting the sheet. When the drops of water dry, they become noticeable defects.

Couching the Sheet

Couching involves transferring the sheet of paper from the mould to the felt. This is done on a couching bench that can be either completely flat or slightly convex. To begin this process, position the mould so that it is lined up with the edge of the felt. Rest the mould on the edge of the felt, and allow it drop onto the bench naturally. Next, press on the mould so that the paper sticks to the felt. Lift it up carefully by one edge, reversing the movement used to place the sheet on the felt. (If you lift the frame all at once, you run the risk of damaging the paper.)

After couching the sheet, place a new felt on top and repeat the operation. When you couch the felt, make sure that there aren't any wrinkles that might leave creases in the paper that would be impossible to remove. Defective felts can also produce creases. The sheets must be centered on the felts, and this is why you should tentatively line them up at the start of each couching. You can couch as many sheets as you desire, depending on the type of felt and the production level that you want. The thickness of the felts also influence how high you make the pile, and whether it becomes uneven or not. Kitchen towels, for instance, are soft and very absorbent, and can result in an uneven pile. When couching onto a tall pile, hold the corners of the mould carefully so that the sheet doesn't rip.

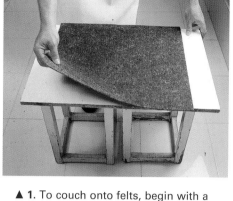

▲ **1.** To couch onto felts, begin with a smooth work surface and a felt. (Many people prefer to start with a pair of felts to provide a better cushion.)

▲ **2.** Gently place the mould and rest it on one of its sides so that the paper will be centered on the felt.

▲ **3.** Allow the mould to tip over naturally, neither slowly nor quickly, as if you were closing a door.

▲ **4.** Firmly push the mould onto the felt so that the paper adheres to it.

▶ **5.** Lift the form by reversing the movement used to place it, as if you're opening a door.

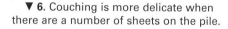

▼ **6.** Couching is more delicate when there are a number of sheets on the pile.

▲ **7.** You'll need to hold the mould carefully by the corners as the pile grows higher.

▶ **8.** As the pile grows, exert less pressure to couch each sheet—as if you're guiding rather than pressing.

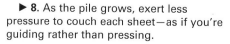

Pressing

When the pile is ready for pressing, cover the last sheet of paper with a large felt or piece of cloth folded in quarters. Immediately thereafter, load the press by putting all the material into place to be pressed—as many felts and sheets as desired—up to the limits determined by the height of the press.

Add a couple of plates at the top of the pile to spread out the pressure, and add planks to fill in space if needed. (Some shops use thick, wedge-shaped pieces of wood that can be fitted together to form a single piece.)

When the pile is properly centered on the press, begin pressing— a quick operation intended to eliminate the water. When no more liquid comes out, back off the pressure and take the pile out. (At this point, the paper doesn't need to be kept under pressure for very long.)

◄ **1.** Prepare the press carefully by centering the pile and placing a large plate on top. (An old-style wheel press is being used here.)

▼ **2.** Place thicker planks on top of the plate to spread out the pressure and take up room in the press, if needed.

▼ **3.** Turn the wheel. The first turn eliminates most of the water.

► **4.** Finish off the process by giving an extra twist to the wheel until it's clear that no more water is coming out of the pile.

Laying

Laying consists of separating the paper from the felt. This is a delicate operation, since some fibers have a very fine texture. In addition, if the paper isn't pressed thoroughly, it may be impossible to peel off the paper without tearing it. (If that happens, it's better to allow it dry on the felt and separate it afterwards.) If the separation isn't done very carefully, the paper may have pinch marks or rips. Begin by using your fingers to lift up a corner and allow it to rest on the broad part of your hand. Repeat the operation using the other hand, and when both corners are free from the felt, continue pulling up the paper. Proceed carefully, because the paper will still be damp, and you need to avoid tearing it.

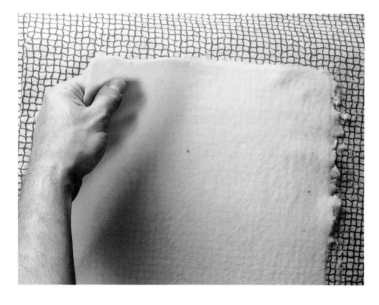

◄ **1.** To lift the sheet, first detach one corner before you lift the other one.

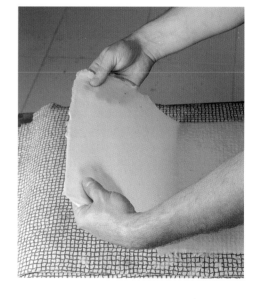

► **2.** When the two corners are free from the felt, carefully pull the paper. (It's preferable to perform this operation on the shorter side of the paper.)

Drying

Next, hang the paper on a clothesline to dry, or dry it between cardboard if you're using the forced drying method. The most common defects that you can cause when hanging it are pinching the sheet with your fingernails and ripping it when taking it from the clothesline.

After lifting the sheet, and while the paper is drying, brush off the felts to remove any pulp remnants. If you wait a long time, it will be more difficult to do, since moisture makes it easier to clean the felts without applying a lot of pressure. (If several people are making the paper, the felts are cleaned as the sheets are being lifted and set out to dry.)

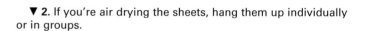

▼ **1.** If hanging the sheets out to air dry, use a piece of perfectly clean, folded thick paper to keep the clothespins from damaging the paper.

▼ **2.** If you're air drying the sheets, hang them up individually or in groups.

▲ If it's difficult to separate the sheet from the felt, hang it along with it.

▲ If you're using a forced air drying system, place the sheets as close as possible to the fan.

◄ **1.** In cases where differences in the thickness of the paper complicate matters of separating the paper from the felt—as with these sheets produced using a collage of colored pulps—allow the papers to dry on the felts.

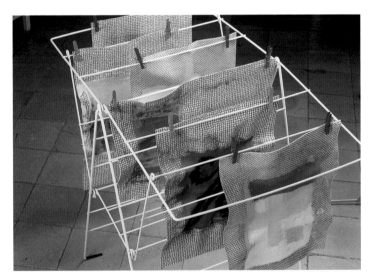

▲ **2.** To remove a sheet that's dried on the felt, stretch the felt to release it from the paper.

▲ **3.** After stretching the felt in one direction, repeat the operation in the other.

► **4.** After stretching the felt along its width and length, it's easier to peel the dried sheet off.

Dry Pressing

Dried paper has been subjected to a certain amount of movement caused by air currents and fiber contractions because the sheets dry from the outside in. Most of the moisture is concentrated in the center. If needed, press the paper again to get rid of any waves and wrinkles. This pressing should last one or two days at high pressure to readjust the fibers. If desired, you can give the sheets a satin finish by pressing them in a small press, or in the original press between two metal plates. A household iron will also produce a satiny effect.

◀ **1.** Often the papers will become very wrinkled as they dry. Before pressing them, carefully straighten them out.

▲ **2.** Ready the pile by alternating the plates and sheets, making sure that they are arranged in such a way that the pressure will be spread out effectively.

▲ **3.** Press them under the highest pressure possible.

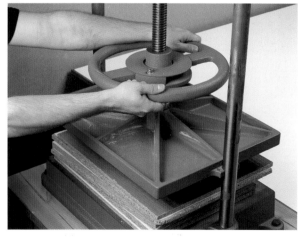

▲ **4.** After the paper has been pressed for two or three days, release the press and remove the paper.

▶ **5.** Your paper will be nice and smooth.

During the papermaking process errors are sometimes made that create defects in the paper. The defective sheets are called culls. Some of them can be repaired by smoothing them out with a strip of wood that is used to iron out the wrinkles and even out abrasions in the dried paper. Defective sheets can also be recycled.

Some of the most common defects are:

- Uneven thickness in the sheet
- Creases or wrinkles that are impossible to eliminate from sheets caused by defects in the felts or because they haven't been placed properly on the laying bench. Pieces of paper with excessively thick deckle edges caused by using too much pulp are in this same category.
- Sheets on which one end is thicker than the other
- Paper with blotches caused by improper external sizing
- Lumps caused by pulp not being distributed properly during fabrication
- Rope marks that cause wrinkles in the sheets when they rub against the drying clothesline
- Crustiness in the paper due to improper distribution of sizing. (This defect used to be caused by too much heat in the drying loft.) This defect is also caused by too little sizing that causes ink to run on the paper.
- Finger marks left on the sheet when it is laid carelessly.
- Marks on the paper caused by droplets of water that fall when the deckle is removed from the mould, or drips from a damp mould on paper that has been laid on a felt. This is the most common defect in handmade paper.
- Sheets of paper with one bad and one usable half
- Pinches in the sheet caused by handling the paper when it is put up to dry
- Tears in the paper caused by hanging it on a rope to dry

- Excess deckle edges on the sheet caused by some defect in the deckle or a failure to fit the deckle to the mould properly
- Excessive sheen caused by polishing the paper too much, resulting in sheets that don't take ink well
- A difference in thickness between the ends of the paper because the pulp wasn't evenly distributed
- Small impurities in the sheet of paper resulting from a dirty vat or mould, or from debris that fell into the pulp during the process
- Corners of sheets with uneven satin finishes

▲ "Papermaker's tears" or drops of water that fall onto the paper when the deckle is removed from the mould

▲ Lumps in the sheet caused by the pulp in the vat not being stirred thoroughly

▲ Paper will not be uniform if the mould is shaken too vigorously or for too long while the sheet is being formed.

▲ Pinching the moist paper as it's being laid can result in tears and marks.

◄ High-pressure water is used to clean the mould and eliminate debris.

▲ If the sheet isn't properly formed and is already on the felt to dry, don't try to scrape the pulp off the felt with your hands because it will cause hard lumps that show up in the next sheet.

▲ To return the pulp from the mould to the vat when the sheet isn't well-formed, hold the mould close enough to the water to moisten the pulp and allow it to fall back in without forming lumps.

▶ Don't touch the sheet with your hands before you return the mould to the vat.

▼ If the sheet is laid on a felt, hold the edges of the felt, and allow the pulp to drop into the water in the vat without touching it to keep lumps from forming.

▼ Every time pulp is returned to the vat, it has to be mixed in thoroughly to break down any lumps that may have formed.

One advantage of paper is its ability to be recycled. After a piece of paper has been used, it can be shredded and washed to isolate its fibers again. Today, nearly three-quarters of the paper that's produced contains recycled paper. This figure shows real progress on the ecological front, since this practice slows down deforestation.

Since the process of removing the fibers and washing them entails a certain loss in fiber quality, it's a good idea—at least with handmade papers—not to recycle paper more than once. The resulting quality can be very poor. Papermaking artisans usually recycle papers by mixing them with purer pulps to create papers that have a particular visual appeal.

In practice, the process is very simple. First, weigh the paper to calculate the final production. (It's a good idea not to exceed an amount of 50% recycled industrial paper in the mix to maintain good quality in the final pulp.) Next, rip the paper by hand to make it easier for the water to penetrate it. (The size of the pieces should be similar to produce a consistent texture in the pulp. The larger the pieces, the less thorough the drainage in the formation of the sheet.) Then, soak the shreds for several hours and grind them up to the desired degree. Pigment can be added to the final pulp to improve its appearance and personalize the paper.

▲ The addition of cut-up pieces of newsprint, or confetti, contribute more to appearance than function.

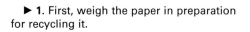

► **1.** First, weigh the paper in preparation for recycling it.

▼ **2.** In order for the water to penetrate the paper thoroughly, tear it by hand into small pieces. Avoid cutting it with scissors or a hobby knife, since a clean cut deters hydration.

▼ **3.** Soak the torn pieces of magazine for two hours.

▼ **4.** You can use a hand-held blender to grind up the paper, but avoid burning out the motor or submerging it too deep, since you could get electrocuted. If you need to recycle a large quantity of paper, use a Hollander beater.

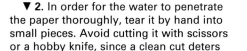

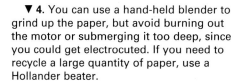

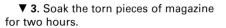

◄ 5. The length of time needed to grind the pulp depends on how visible you want the fragments of recycled paper to be.

◄ 6. Pour the pulp into the vat. Here, there is a portion of colored, high-quality pulp to counter the neutral gray of the printer's ink in the recycled paper.

▼ 7. Because of the small, scale-like fragments of paper, recycled paper drains more slowly in the mould.

▼ 8. Recycled pulp up close

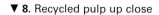

◄ 9. Add paper confetti to heighten the chromatic effect.

◄ 10. The attractive end result of adding confetti to recycled paper

Making Decorative Papers Containing Added Materials

There are two ways to incorporate different materials into paper. One way is to mix in other substances or objects with the fiber, without any order or composition; and the other is to arrange these objects in some predetermined pattern. In both cases, the information covered in the next few pages concerns the kinds of things which blend well into the pulp and those which don't, the fact that the materials should be moistened before adding them, and so forth. As a first example, you'll see how to add flowers to paper, and then you'll see how to add materials with a greater volume.

▲ Marigold petals are a good choice to include in paper because they tend to integrate well with the pulp.

▲ Some green cypress leaves, mint, and maidenhair fern provide a bit of color contrast to paper.

▲ **1.** Soak the petals and leaves for two hours before adding them to the vat.

▲ **2.** Make certain that the flowers mix in with the pulp without floating to integrate them well.

◄ **3.** Petals that have been mixed in rise to the top of the pulp. If you use a mould with a very fine mesh that allows the pulp to drain slowly, you can choose to arrange the flowers.

▼ **4.** A petal or two will usually stay on top of the sheet and fall off during the drying process. To avoid this, try to push it back down into the sheet as it is forming.

When you incorporate materials according to a preconceived design, place them on the paper in the mould or felt. Then add some highly diluted pulp on top so that it mixes with the paper during pressing. If you're using a mould with a very fine mesh and a high deckle, you can place these materials in the mould while the pulp is draining, allowing time to push them down into the pulp a bit. Add some diluted pulp to the areas that seem as though they need better integration between the pulp and materials.

▲ Papers containing flowers

► **5.** To add flowers and leaves to the top of the sheet, place them on the pulp in the mould.

◄ **6.** Add a light coat of pulp from the vat on top of the materials so that they integrate with the sheet. Don't cover them entirely, or the design won't be visible when the paper dries.

► **7.** When you use a mould with a very fine mesh and a high deckle, the drainage is slow, which allows you to take your time and move the parts of your design around.

◄ An album made with paper containing flowers

▼ An assortment of papers containing added materials

▼ A container made from heavy paper that contains flowers

Another method of adding materials is to couch the sheet onto a felt, and then place the materials on the sheet before covering them with another sheet of pulp, like a sandwich. With this method, the objects give the paper body and volume, and the shapes of the materials create a relief. Materials that work well are straw, reeds, metal objects, and textiles. After the object is covered with the second sheet, it will be partially visible because the pulp will thin out in certain places.

▲ **1.** Here, the first layer of paper is placed onto a piece of burlap used as the felt to give it a textured finish. Strings are then arranged on this layer.

▶ **2.** Line up the second layer of pulp on the mould so that it will fit on the first.

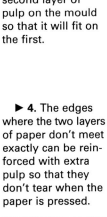

▶ **4.** The edges where the two layers of paper don't meet exactly can be reinforced with extra pulp so that they don't tear when the paper is pressed.

▲ **3.** The second layer of paper reveals the shape of the strings, even though it hasn't yet bonded to the first.

▲ **5.** If desired, some of the material can be revealed by scraping away some fragments of paper.

▶ **6.** The final result after drying is a highly textural paper.

Basic Color Techniques

One of the most attractive techniques for small-scale papermaking is making paper with added color. Visual artists who work with paper as a medium for their creations often use colored pulp as a raw material. To maximize the success of coloring, make up a good deal of base white pulp, and store it before beginning so that there's plenty with which to work. The base pulp should be highly refined, without any lumps, and fairly diluted so it can be applied using applicator bottles.

The colors that you use depend entirely on your preference as well as the palette of colors available, if you're using pigments. If you use colors made from natural materials, you won't get many bright colors. Artificial colorings are good for making brighter colors, and the pulp will naturally modify the colors somewhat so that they lean in the direction of pastel tones.

Colored pulp tends to change in value when it dries, becoming lighter and more luminous, a change that should be considered when making a project. When you're trying to match a specific color of paper or pulp, begin with the dried version rather than liquid, because it can be very deceiving. Dried pulp—almost always in the form of paper—should be ground up in water so that it can be matched as it looks at this stage.

Similarly, if you're adding color to pulp inside the vat, you'll need to see what the color looks like when the paper dries. To do this, take out some of the wet pulp, squeeze it thoroughly between your fingers to remove the water, and do a preliminary check of the color. You can complete this test by drying the piece of pulp with a hair drier.

▲ **1.** If you want to create a certain color, start with a piece of dry paper of that basic color. Wet it and beat it to see its color when wet, which is always darker.

▲ **2.** Add pigment to the vat until you get the desired shade.

▼ **3.** After adding the pigment to the pulp, stir it to mix it thoroughly. Some pigments are toxic, so it's not a good idea to put your hands into the vat until the pigments are completely dissolved in the fibers.

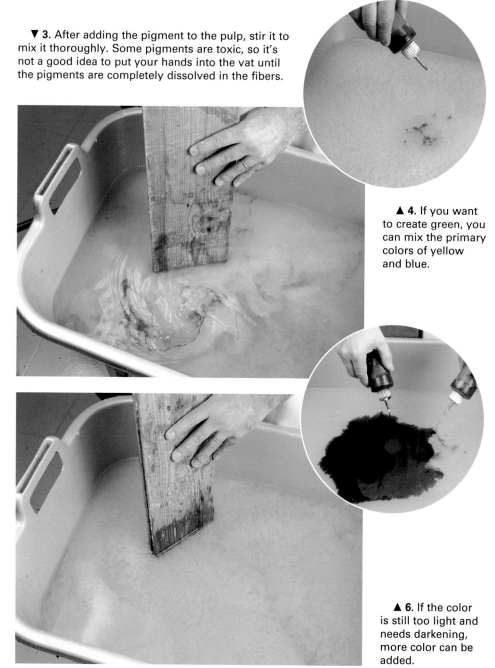

▲ **4.** If you want to create green, you can mix the primary colors of yellow and blue.

▲ **6.** If the color is still too light and needs darkening, more color can be added.

► **5.** Once you've added the two pigments, mix the pulp again.

► **7.** The new mixture is the desired color.

► **8.** To check the color, make a sample sheet, since the color always looks darker in the vat.

▲ **9.** If you want to know what the color will look like when it's dry, scrape off a little pulp from the mould, and squeeze it tightly with your fingers.

▲ **10.** Squeezing the pulp removes the water and lightens the color.

▲ **11.** Check the color, and fine-tune the process by drying the pulp with a hair dryer.

▼ **1.** To create a color for a specific application, prepare the mix in a small applicator bottle filled with pulp. Here, we're making a vermilion from a base of yellow pulp.

▼ **2.** A small drop of red is all it takes to brighten the color and change it to vermilion.

▼ **3.** A couple of shakes colors the fibers perfectly.

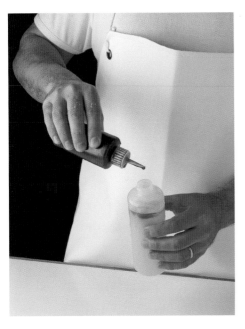

Painting with Pulp

Never apply pigment directly to a moist piece of paper in the mould or on a felt, unless you're using a soft, natural pigments (such as those derived from tea, coffee, or onion skin). Otherwise, using artificial pigments full strength can stain the felts, the dye will run on the paper, and be difficult to control. (Keep in mind that artificial coloring is very powerful and hard to clean up except with bleach.)

To control the intensity, always add pigment drop by drop to diluted pulp that you've put in an applicator bottle. Shake the bottle, and check the degree of color by applying a small line to a piece of blotting paper or unrefined pulp. If the pulp doesn't have too much pigment, the damp halo of the line will be transparent. If there is too much pigment, the halo will clearly show the presence of dye.

A simple color technique involves painting with small plastic applicator bottles with pour spouts for the pulp. Keep in mind that the pulp must be highly refined so it doesn't form lumps that plug the applicator spout. As long as you don't squeeze too tightly, this method works well for making lines. If the pulp comes out with too much force, it can ruin the sheet.

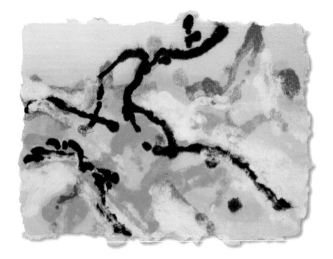

◄ This design was done by combining lines of colored pulp.

▼ 1. Prepare a palette of colors you intend to work with, coloring the pulp as described on the previous page.

◄ 2. If the right proportion exists between pulp and pigment, and it is applied to blotting paper or unrefined pulp, the moist halo will be clear. But if there is too much pigment, the halo will contain the color, looking like ink that has run. In that case, dilute the mixture with more pulp before using it.

▼ 3. Couch a sheet onto a felt. It will act as a canvas for your colored design.

▼ 4. Use one color to make the first strokes.

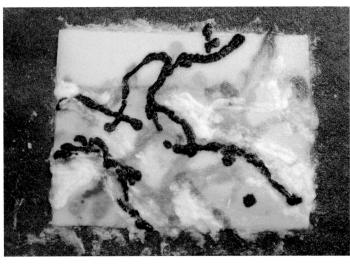

▲ **5.** Add more color to the design with other colors.

▲ **6.** When you're happy with the design, it is complete except for pressing and drying.

▲ ▲ ▼ ▼ Four designs incorporating colored pulp

Gradations and Masking

You can also create paper of two colors. There are a couple of different ways to do this: First, make a partial sheet with an initial color before placing a second sheet on top of it that completes it.

The second is ideal for creating background colors or two colors on both sides of the sheet, and it uses a color gradation. Two sheets are made by overlapping several colored sheets and blending the borders between the colors. You can make variations based on two-colored paper; for example, the color

of the first layer can appear in the second one if you use a mask to block off a portion of the latter as the sheet is being formed. You'll need the assistance of another person to hold the mask in the mould as it's placed in the vat. The mask is usually made of some solid material and should be firmly pressed against the mould so that no fibers get underneath it. When the mould is removed from the vat, it drains normally, and the mask is lifted only when all the water has run off. When it is removed, the silhouette of the form remains, acting as a window through which the paper underneath it is visible.

▲ **1.** Make the first layer by inserting the mould halfway into the vat full of colored pulp.

▲ **2.** Couch this fragment onto a felt.

◄ **3.** When couching the second color, carefully calculate how the two fragments fit together so that a sheet is formed with parallel edges.

▼ **4.** The completed gradation

▼ **5.** Once it has dried, the paper has a gradation from green to yellow.

▲ **1.** You can use something as simple as a hand as a mask. Use the help of another person while you push the mould into the vat.

▲ **2.** After the mould is removed from the vat, hold it in place while the water drains, keeping the hand (or other mask) in place the whole time.

▼ **3.** Once the drainage is complete, lift off the mask, leaving a negative of the shape.

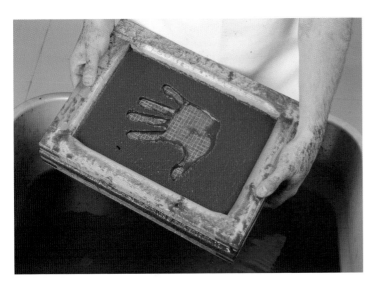

▲ **4.** Line up the mould with another sheet of paper of a contrasting color. Center the image.

◄ **5.** The final result as it looks when it is still wet

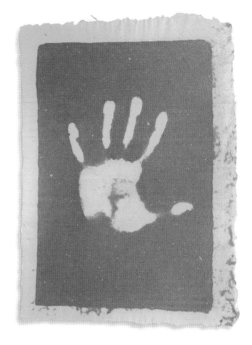

► **6.** The paper after it dries

Collage

A final color technique that's interesting to use in papermaking is collage. This term, which comes from French, means "to glue." Making a collage normally involves gluing papers onto a background. In papermaking, the collage is made as the sheet of paper is being formed, so no glue is needed. The fibers of each paper bond with the ones beneath it. Water is the medium that connects the fibers of each layer.

The layers of paper added to the composition can change dramatically depending on how the paper is handled in the mould. Layers can be added as you wish, combining them with other techniques we've already explored (such as adding lines, blocking off areas, and using color gradations). A mould with no support strips on the back is used for paper collage, since you need to see the image through the mesh to work with the pulp without interference.

▲ This work by Manel Lancis combines a collage of colored pulp with a transferred image.

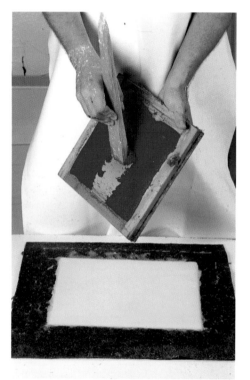

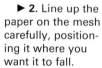 **1.** To make a collage, begin with a couched background paper. You can scrape away part of the colored paper on the mesh to leave a section to collage onto the background.

▶ **2.** Line up the paper on the mesh carefully, positioning it where you want it to fall.

▼ **3.** Press lightly on the back of the mould so that the second layer adheres to the background.

◀ **4.** You can texture the second layer with your hand to remove a portion of the pulp.

► **5.** Add the textured layer on top of the sheet, allowing it to overlap.

► **6.** Apply a final layer of pulp to make up a three-layer collage. (You can add as many layers as you wish.)

► **7.** This is the finished collage to which transfers can be added.

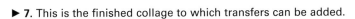

▼ You can mix and transfer color to the mould from a measuring cup as an alternative to using a vat.

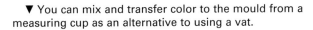

▼ To create a layer made from colored pulp made in a vat, you can insert the mould partially into the vat to pick up the pulp.

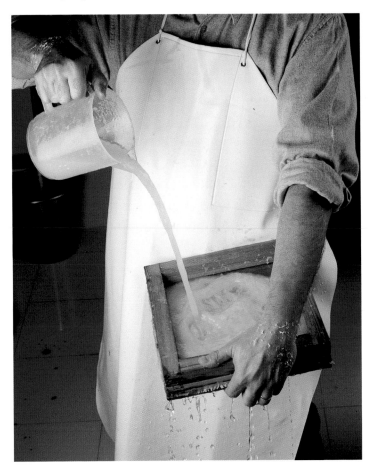

*T*HE plant world is the main source of paper fibers, but plants differ in how much fiber they provide and the quality of the fiber. Some plants yield excellent fiber from their stems, others have good fiber in their bark, and some have good fiber in their fruits and leaves. But every one of these types of fiber are different: long, short, broad, narrow, stiff, strong, porous, and so forth. People who make paper need to know the characteristics of the main fibers used in papermaking, since these are the raw materials of their work. If, for instance, you want to make a paper with great mechanical resistance, you'll need long fibers that can be refined, such as fibers from linen and conifers; and, if you want to produce a specific volume, you'll need to use cotton. In order to make paper with a good surface finish and volume that takes print well, esparto grass or the short fibers of leafy trees such as eucalyptus and birch can be used. Straw from grains or bagasse can be used to make stiff paper. Every type of fiber has its own properties, and they can change during the process of making the pulp. Combinations of fibers make it possible to design a certain type of paper with precision.

Fibers

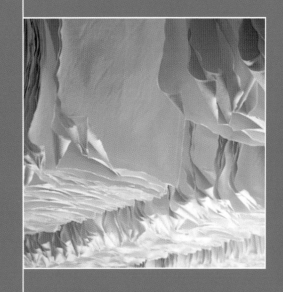

The Origin of Fibers

▲ Even though bamboo is a type of grass, it grows to a great height. The immature plants tend to contain the greatest concentration of cellulose in their stalks.

► The straw from grains produces a high-quality fiber that gives the paper strength and stiffness.

► On a global scale, wood supplies 91% of the pulp for paper production.

The Chemical Composition of Fiber

The chemical composition of fiber varies from plant to plant. There are also variations inside different parts of the plant, even within a single fibrous wall. Nevertheless, if thinking broadly, we can divide plant compositions into two major classifications: the main components (cellulose, hemicellulose, and lignin) and the minor components (mineral and protein substances, fatty acids, resinous acids, phenols, and so forth).

Cellulose

Cellulose is like the skeleton of plant cells. Its main characteristic is that it is hygroscopic: it absorbs water and swells. Even though it is insoluble in water, it is a polymer that will decompose. Decomposition can happen through several means: water, oxygen, alkaline pH, heat, or microorganisms.

Not all plants have the same percentage of cellulose, as these examples will demonstrate:

FIBER	% OF CELLULOSE
Cotton linter	96
Hemp	> 60
Linen	> 60
Abaca	53–64
Sisal	53–64
Bamboo	> 50
Conifers	40–45
Leafy Plants	38–49
Esparto Grass	42
Bagasse	40–43
Reeds	33–43
Corn	33–43
Straw from grain	33–38
Rice straw	28–36

◄ In paper, the fibers retain the same characteristics they had in their plant source. Thus, just as wood swells and changes shape with moisture, so does paper. That's why these piles of paper are stored with a weight on top, to keep them from changing shape due to the relative humidity in the air.

Hemicellulose

Hemicellulose is a part of the plant that is any of several polysaccharides that are more complex than a sugar and less complex than cellulose. It acts as a substrate for the cellulose microfibrils in the cell wall and is very hydrophilic. It promotes fibrillation inside the fibers, improves their flexibility, facilitates refining, and increases the ability of the fibers to bond together. Hemicellulose breaks down more easily than cellulose.

Lignin

Lignin is an amorphous polymer that acts as a natural binder and support for the cellulose fibers of woody plants. It binds by building up the intercellular spaces and the fiber walls. Lignin provides rigidity, cohesion, and resistance to compression. It also protects against the invasion of microbes. Lignin is hydrophobic: it does not absorb water, and it interferes with the swelling and refining of the fiber. It softens only with temperature, which facilitates the extraction of fibers—the method that is used in to mechanically refine pulps. The proportion of lignin is different in each plant. For instance, conifers have a greater percentage than leafy plants, which in turn have more than straw.

Extracts

Extracts are substances (such as minerals, proteins, pectic acids, fatty acids, resinous acids, and phenols) that are found in small quantities in fibers. They have no influence on the structure of the fiber, but affect how the paper pulp is produced. They contribute color and odor to the wood and are the source of secondary products such as turpentine. Their presence lends color to the paper pulp. If they are present in significant quantity (as in old pine trees, where they compose seven to 24% of the core of the wood), they make the material inappropriate for use in paper pulp.

The Structure of the Fiber Wall

The microscopic structure of cellulose has a system of capillary walls in the tissue, composed of macrofibrils and microfibrils.

A cross section of a fiber reveals the following layers:

- The medial layer, which establishes the bonds with adjacent fibers
- The primary wall, which is torn during refining, thereby making it easier for the water to gain access to the innermost layers of fiber and the external fibrillation
- The secondary wall that has three layers formed by nearly parallel microfibrils. The lignin and hemicellulose are located among these layers.

The angle formed by the microfibrils with respect to the longitudinal axis of the fiber plays a decisive role in the physical behavior of the fiber; it's also crucial in establishing the characteristics of a sheet of paper.

- The hollow inside is known as the lumen.

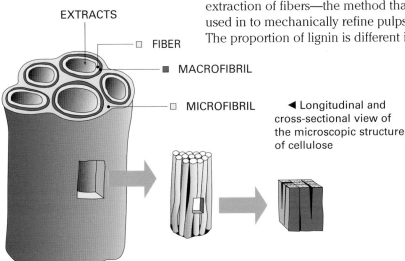

EXTRACTS

FIBER

MACROFIBRIL

MICROFIBRIL

◄ Longitudinal and cross-sectional view of the microscopic structure of cellulose

Properties of Cellulose Fibers

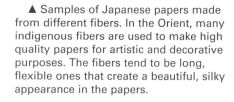

The following collective properties of plant cellulose fibers make them the ideal component for paper:

- High resistance to traction
- Adaptability (flexibility and ability to conform)
- Retention of shape
- Insolubility in water
- Hydrophilic quality
- Broad range of dimensions
- Ability to establish bonds among fibers
- Ability to retain additives that modify their properties
- Chemical stability
- Relative whiteness
- Renewable resource

In theory, it seems as though all plants could be used for making paper, but in practice, it doesn't work that way. Not all of them have the right percentage of fiber, and others don't exist in the quantities needed.

When we think of plant fibers for paper, we refer to the ones that come from trees, but it's only recently that trees have been used as a source of fiber. Up until the middle of the nineteenth century, non-wood fibers were used exclusively to make paper. Currently, trees account for 91% of the worldwide supply of materials for paper. In some developed countries that figure is 99%. This high percentage is due to the amount of fiber available in trees, the abundance and variety of species, and the low costs of storage and treatment. All these factors make wood fibers economical for producing high-quality paper.

Fibers from sources other than wood have recently come back into use. In some developing countries, these are the main, and sometimes the only, source of fibers for making indigenous papers. The use of non-wood fibers inhibits the uncontrolled deforestation that disrupts the ecological balance of the planet, since most of these fibers come from agricultural operations (such as straw, bagasse, and so forth). Non-wood fibers make it possible to manufacture paper of such quality and variety that wood isn't needed.

Animal fibers (wool, fur, and silk), minerals (asbestos, glass, and metals), and synthetics (rayon, nylon, polypropylene, and others), are mixed with plant fibers to enhance the stability of the sheet, water repellence, strength, and other special characteristics.

▲ Samples of Japanese papers made from different fibers. In the Orient, many indigenous fibers are used to make high quality papers for artistic and decorative purposes. The fibers tend to be long, flexible ones that create a beautiful, silky appearance in the papers.

▼ Pine fiber treated with sulfate (kraft) was used in making this paper sack that contains 55 pounds (25 kg) of plaster. The long, rigid fiber from conifers is ideal for making strong papers.

▼ Sample of papers made by hand using various fibers from wild plants. In some cases, it's very difficult to remove the natural color or refine the fibers adequately. Every species requires the right kind of processing.

▼ Commercial papers in sketchbook form for use in art. Different uses of paper call for different fibers.

▲ Paper used for currency is one of the toughest made. Normally it contains a high percentage of linen fiber.

▲ Fiber for cigarette papers commonly comes from straw or linen tow that's grown for seeds. This paper is very fine, has some stiffness, and its qualities cause it to burn at an appropriate speed.

The Influence of Fibers on the Paper

The properties of paper pulps and the resulting papers depend directly on the characteristics of the fibers. The processing to which they are subjected has a major impact on their morphology and strength; in general, though, most of the lignin has to be removed to make the best pulp. The most common reasons for choosing one fiber over another are its characteristics of length, diameter, wall thickness, size of lumen, flexibility, rigidity, and so forth.

The relationship of the length to the diameter of the fiber and the thickness of its wall is the first criterion used to evaluate good pulp. Papers made with fairly long fibers are always the strongest, tear less easily, and are more porous. Long wood fibers are broader and thicker than short ones, and therefore, more rigid. Although a certain length is good for making paper, length becomes a negative aspect if the fibers are too long, since it becomes difficult to produce a uniform sheet.

The breadth of the fibers varies a lot among different species. Wood fibers tend to be thicker than non-wood fibers, which are a good deal thinner. The thinner the fibers, the greater their ability to bond with one another. The thickness of the cell wall and the mass of the fibers have an influence on the paper, affecting porosity, smoothness, strength, and opacity. It stands to reason that sheets made from thick fibers have fewer fibers and bonds per gram than those made from fine, thin fibers. Fibers of different lengths and thicknesses are almost always combined to produce sheets that are stable and suited to many applications.

LENGTH AND WIDTH OF CERTAIN FIBERS (in millimeters)		
Fiber Length	**Width**	**Length/width ratio**
• *Non-wood*		
Esparto grass 1.5	0.011	136
Straw from grains 1.5	0.013	115
Bagasse 1.7	0.020	85
Bamboo 2.7	0.014	190
Abaca 1.8–6.2	0.011–0.018	254
Sisal 1.3–2.7	0.019–0.037	35–142
Cotton (linter) 2.0–12	0.020	100–600
Cotton (fiber) 12–50	0.09–0.023	1000–4000
Linen 10–36	0.011–0.020	1100–1200
Ramie 40–200	0.045	880–4500
• *Wood*		
Pine 2.0–3.0	0.022–0.050	60–90
Spruce 3.1–3.5	0.019–0.050	70–160
Poplar 1.5	0.025	60

Wood Pulps

Conifer Pulp

Conifer trees, of which there are more than 550 species, are the main source of long wood fibers. They grow in thick forests, and their trunks are long and straight. Their wood contains a high concentration of fibers. More than a third of the world's forests are made up of conifers (such as pine, spruce, hemlock, and cypress), and three-quarters of the wood fiber used for industrial purposes comes from this source.

Morphology of Conifer Fibers

Conifer fibers are known technically as longitudinal tracheids. Ninety percent of the total volume of a conifer trunk, minus the bark, is composed of longitudinal tracheids. The remaining ten percent is made up of other, smaller fibers that are lost during the processes of washing and purifying the pulp.

Longitudinal tracheids are elongated fibers (mostly between 3 and 5 mm and up to 7 mm) that are also thick (between 0.03 and 0.06 mm), depending on the species. These fibers are lignified, strong, blunt or pointed, depending on how they are observed. Besides their length, they have multiple perforations—thin spots in the walls connecting the internal cavity of the fiber (the lumen) to the neighboring fiber.

▶ Wood fibers are used in large paper mills.

Treatment and Use of Conifer Pulp

Conifer pulp is widely used, and the morphological characteristics of conifer fibers make them an ideal source for making paper. These papers are uniform and strong. Adding conifer fiber to pulp mixtures improves the strength of the finished paper, especially kraft pulps.

Conifer pulps can be produced by chemical or mechanical means. The choice of method depends on the species of tree, the wood color, and the resin content. Woods containing a greater percentage of resins and oily substances are best processed in an alkaline medium (with a high pH) such as sodium sulfate.

Spruce and poplar, which have light, delicate wood with little resin, can be used to make all types of pulp (chemical—alkaline or acid, semi-chemical, or mechanical).

Pines are almost always processed with sodium sulfate to make kraft pulp because their high resin content, their color, and their hardness are not appropriate to either the mechanical process or chemical treatments using acid sulfite (pH around 2).

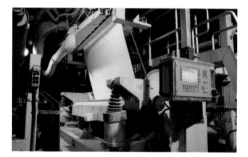

▲ Conifers fulfill the need for three-quarters of the wood used to make paper.

Pulp from Leafy Trees

Leafy species are also known as deciduous trees or hardwoods. Among all the leafy trees, the ones most commonly used for paper pulp are birch, eucalyptus, beech, linden, poplar, and liquidambar, although there are also many other useful species (ash, oak, walnut, elm, maple, etc.).

Traditionally, fibers from these trees have been rejected for paper production, and used only as filler for papers that don't need much strength, since the fibers are short and the pulp less uniform. Another disadvantage of these woods are their porosity in water, which made them unsuited for transport by river. They are harder to use in a mechanical pulping process because their trunks are not always straight, the bark is more difficult to remove, they have small, knotty branches, and the these trees are sometimes too dense.

However, two-thirds of the forested area in the world are made up of deciduous trees. As a result, new technology has been developed for improving the pulp obtained from them to help meet the growing demand for wood pulp.

Morphology of Fibers from Deciduous Trees

Whereas conifers have a very uniform and homogeneous fiber composition (mostly containing longitudinal tracheids), the composition in deciduous trees is very heterogeneous, and there are many types of cells relating to

◀ Conifer fibers are long, strong, and recognizable under the microscope by the perforations in their walls.

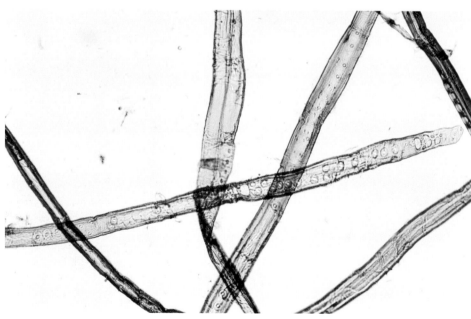

▲ Eucalyptus fibers under the microscope. Note the details of the vascular tissue.

▶ Papers made from wood fibers. The composition is a mixture of conifer and eucalyptus.

different purposes (mechanical support, conduction, or intermediate functions). The main types of cells found in this type of pulp are libriform fibers, tracheid fibers, vascular tracheids and vasicentric tracheids, parenchyma cells and vessels. The first two cells are found in all species and simply referred to as "fibers." They make up about 50% of the volume of the tree.

Libriform fibers have thick walls, pointed ends, and very small perforations that are difficult to see. Tracheid fibers have a wall thickness that varies but is thicker than libriform fibers; they are broader and have a larger lumen. Their ends are rounded, and they have rings around their perforations like conifers, although they are smaller and less numerous. The average length of these fibers falls between 0.75 and 1.5 mm. Their average breadth varies between 0.01 and 0.02 mm. (Of course, there are always species that deviate from these averages.)

The vessels of these trees are made up of tubular-shaped cells that appear to be connected to one another to form longitudinal vascular tissue. They represent ten to 30% of the plant's volume.

▼ Poplar fibers

The walls of the vessels exhibit perforations at the ends as well as along the sides. They make up groups of reticular-shaped, scalariform, simple (parallel), or circular orifices. Also, they have a fairly pronounced appendix at each end, and in some species, there are spiral thickenings along the walls. Their length varies between 0.4 and 0.8 mm, and the diameter is between 0.02 and 0.5 mm. Under a microscope, the vessels are used to identify the various types of hardwood pulp, since they vary from one species to another.

Parenchyma cells make up 20% of the plant's total volume. These are small, brick-shaped cells that contain extracts such as gums, oils, resins, starch, latex, and others. Depending on the ultimate quality desired in the pulp, it's a good idea to remove these cells.

The Most Commonly Used Hardwood Pulps

- Eucalyptus: This tree is native to Australia, with around 600 species spread throughout the world. The prolific nature and the quality of the wood makes it one of the most commonly used hardwood pulps.
- Poplar (of the genus *Populus*): This tree grows very quickly and is easy to grow in moist terrain. The wood is light, homogenous, and easy to process. It's used for high-performance, semi-chemical, raw chemical, and bleached pulps.
- Birch (from the genus *Betula*): This tree has broad applications in the paper-making industry in the Scandinavian countries. The wood is white and homogeneous; it's used for bleached sulfate pulps and in high performance pulps.
- Beech: This tree is used for paper in Europe (*Fagus silvatica* species) and North America (*Fagus grandifolia* species). It has white, hard wood, which makes it unsuited for mechanical pulp. It's fine, however, for sulfate pulps, whether raw or bleached.
- Linden (genus *Tilia*): Like beech, this tree has also has two species. In comparison with other hardwoods, the fibers are much broader, and there are spiral thickenings on the walls of the vascular tissue.
- Liquidambar (*Liquidambar styraciflua* species): This wood is used only in North America. Its vascular vessels are very long (between 1 and 1.4 mm) and narrow.

Characteristics and Uses of Hardwood Pulps

All hardwood pulps are weaker than conifer pulps. Even the ones with more length to their fibers are less resistant to tearing, although their resistance to rupturing and tension is a bit better. Normally, these pulps are used for printing because of the positive surface qualities that they impart to the sheet. Hardwood fibers are added to mixtures to improve sheet formation, smoothness, paper opacity, and body.

Non-wood Pulps

Non-wood fibers include grasses, canes, straws, and other annual plants, accounting for most of the paper production in less developed countries. In some countries, it accounts for the entire pulp production. The production of non-wood fibers is highest in Africa, Asia, and Latin America. Straw from grains and bagasse are the non-wood fibers most commonly used in paper-making. Mexico, China, and Peru are leading producers of bagasse pulp. China, India, Spain, Italy, and Turkey are leaders in the production of straw, with China producing 80% of the worldwide total. India leads in production of bamboo (a grass) pulp.

Non-wood pulps are classified into three major groups based on the origin of the plant:

1. Naturally growing plants: bamboo, cane, esparto grass, and others

2. Agricultural residue: grain straw, sugar cane bagasse, sorghum, cornstalks, cotton stems, etc.

3. Plants cultivated for the purpose: derivatives of stalks (linen, hemp, jute, ramie, crotalaria, kenaf), of leaves (abaca, sisal, henequen), or from seed-pods (cotton fibers and linters). Recycled rags and ropes are considered part of this group, since they come from cultivated fibers such as cotton, linen, hemp, jute, and sisal.

Morphology of Non-wood Fibers

The dimensions of these fibers and their pulps vary greatly. Some non-wood fibers are as long as hardwoods, and some are shorter. They are generally small in diameter. The morphology and typology of these fibers can vary in a single plant, depending on where it's grown. It's possible to categorize the following, grouped according to the type of plant:

Assorted Non-wood Plants

Linen, hemp, ramie, jute, kenaf, kozo, gampi, mitsumata

- Assorted fibers are found in the tissue if these plants that is known as phloem, which is located between the xylem (timber) and the outermost bark. Several types of cells make up the phloem. Some fibers are longer, harder, and thicker than others, depending on how close they are to the bark. Their dimensions vary from fractions of a millimeter to 1½ feet (.5 m). In the case of ramie, their diameter varies between 0.02 and 0.05 mm.

- Short fibers (between .5 and 1 mm long) come from the heartwood when the whole plant is used

Monocotyledons

Wheat, barley, rye oats, rice, sugarcane, common cane, corn, esparto grass, etc.

- These plants have medium-sized fibers of more heterogeneous dimensions than wood, with the exception of bamboo, which is much larger. Among the straws, knots and folds are evident, the ends are pointed, and the size of the lumen varies.

- Sclerotic cells serve as a support for these plants. Their walls are lignified, variable in shape (more or less polyhedral, elongated, in the shape of a stiff rod or similar to fibers). They have a small perforation and a lumen almost as large as the whole cell.

- Parenchymatous cells, also called "tunnels," small and with ends not perforated and round in shape, of variable dimensions and thin walls

- Epidermal cells, also called "combs" because of the peculiar toothed shape of their long sides, very abundant in pulps and of variable dimensions

- Tube-shaped, ringed, or spiral vessels

▲ Harvesting sugar cane in Barbados. The use of straw and stems from many species cultivated for other purposes, such as sugar cane and grains, contributes to a solution to the serious problem of deforestation that is fueled in part by the demand for paper.

▲ Iris paper

▼ Paper made from corn and cotton

▲ *Kozo*, also known as paper mulberry. Along with *mitsumata*, this is the outstanding fiber used in Oriental papers.

▼ Under the microscope, linen fiber shows X-shaped transverse marks known as knots. Under tension these knots disappear, then reappear when the fiber relaxes. Their lumen is very narrow and looks like a thin black line in the center. In hemp, this lumen is broader and more continuous.

▼ Kenaf plant. Bast fibers are extracted from its bark, and the much smaller woody fibers are extracted from its core.

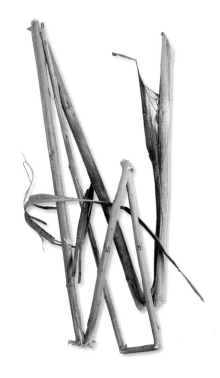

Products from Dicotyledons

Cotton

- Cotton textile fibers. By nature, these fibers are individual units, with a total length between 12 and 33 mm and a diameter between 0.016 and 0.21 mm, separated from the seeds in processing. They are used primarily in the textile industry. They have a flattened, twisted, filament-like shape, depending on how ripe they are. Their lumen takes up half their diameter, and their walls are thin. The relationship between the length and width contributes to the excellent quality of rag paper made with these fibers.
- Cotton linters. These are the short fibers that stick to the seeds when the longer fibers of cotton are removed. Linter pulps are chemically processed to be very smooth. Even though they are smaller, the short fibers look very similar to the long ones. One end is pointed and the other (which is inserted into the seed) is broader.

The Most Commonly Used Non-wood Pulps

Linen (Linum usitatissimum)

This fiber, native to Asia, is cultivated for its seed, oil, and fiber for making textiles and paper. Its bast or soft fiber has a length between 10 and 55 mm, with an average of 28 to 30 mm, and its diameter varies between 0.012 and 0.030 mm. The fiber wall is very thick, with a narrow lumen. Linen has a great capacity for bending and folding and is quick to absorb and shed water, which makes it good for clothing. It is used to make tough, resistant, dense, permanent papers. It is also ideal for fine but durable papers for such things as cigarette paper, Bible paper, and paper money.

Hemp (Cannabis sativa)

Hemp is of Asian origin and is the oldest of the cultivated plants. Today, it is raised in many countries, since it adapts easily to many climates. It is an annual leafy plant that grows up to 16 feet (5 m) tall, although it is harvested for paper when it's about 6 feet (2 m) tall. The stem has a structure that is very similar to linen. From 15 to 20% of the dried stem mass is made up of bast fibers measuring

▲ Mitsumata is an arboreal plant used widely in the Orient that yields very high quality fibers.

between 5 and 55 mm in length and between 0.016 and 0.05 mm in width. In addition, it contains other short, woody fibers similar to hardwoods. Hemp fibers can be separated more easily than linen. It's used for alone or mixed with other pulps to make tough, high-quality papers.

Kenaf (Hibiscus cannabinus)

Widely cultivated in India, the American southeast, and Central America, this plant has roots in ancient Africa and has a name of Persian origin. It grows quickly in the form of slender, tall canes that can reach maturity in 120 days, making it very profitable. The bast fibers of the cortex measure between 2 and 6 mm in length, and the woody fibers of the heartwood of the plant are from 0.5 to 0.7 mm long. These fibers are exceptionally strong in their resistance to tension, explosion, and tearing—comparable to bleached conifer fibers that are made into kraft pulp. The fibers also produce opacity, porosity, and a smooth surface in paper, so they're used for making cigarette papers, bank paper, tea bags, filter paper, newsprint, Bible paper, non-sticking papers, and other papers.

Straw from Grains

Historically, straw from grains, particularly wheat and rye, was one of the main resources for paper before wood fibers came into use. Today, the high yield of the raw pulps are used to make cardboard. Chemical pulps made from bleached straw are used as the main component in quality papers made for writing and printing. Straw paper has a hard rattle and good resistance to rupturing, but not as much to tearing and creasing. It is commonly mixed with other pulps containing long fibers.

Bagasse

Bagasse is a byproduct of sugar cane—a fibrous residue that's left after removing the sucrose in the pith. It's an economical source for pulp, since the pith is removed when the sugar is processed. The fiber is fairly short (between 0.8 and 2.8 mm), but it has a tremendous number of applications such as printing paper, paper napkins, paper towels, paper cups, grease-resistant papers, multi-layered papers, and more.

Esparto Grass

This grass is native to Northern Africa and southern Spain. The usable part of the plant is in its leaves rather than stems. The fibers from esparto are very elastic, short, and narrow (with an average length and width of 1.1 mm and 0.09 mm, respectively). The cross-section of the fibers are cylindrical, with pointed ends and a regular but very narrow lumen.

Cotton

Cotton fibers are a product of the cotton plant and grow on the epidermis of the seeds. Cotton is cultivated in the warm regions of America, Asia, Africa, and Europe. The long fibers are used in making textiles, so they can be recycled for paper; the short fibers (linters) are used directly in making paper. Bleached linters are used for papers that require very pure cellulose, durability, permanence, smoothness, good workmanship, and opacity. They are also used for high quality art, drawing, printing, and writing papers. Mixed with linen, they are used in paper money, filters, and other applications.

Abaca (Mussa textilis)

Abaca (Manila hemp) grows in the Philippines, Indonesia, and Central and South America. Along with sisal, it is considered to be a "hard" fiber because it is thicker and stiffer than the "soft" fibers such as linen and hemp. The length of the fibers varies between 2 and 12 mm; the average is 6 mm. The width is between 0.012 and 0.046 mm. The fibers are cylindrical and of uniform diameter; they are sharp at the ends, with subtle transverse striations.

► Schematic diagram of the cross-section of a cottonseed: short fibers (linters) (1), and long fibers for spinning (2)

Sisal (Agave sisalana)

Originally from Mexico, sisal grows in tropical climates. As with abaca, the fibers are extracted from the plant's thick, broad leaves that end in a hard spine. This fiber is as hard as abaca and is used primarily for cordage and weavings. In Mexico, however, it has been used for many centuries in making paper. The fibers are some 3 mm long on average and 0.02 mm thick. They are cylindrical, with a broad, rounded end and fine striations on the fiber walls.

▲ Straw fibers under a microscope

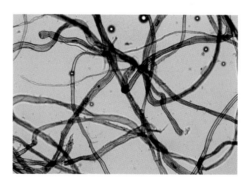

▲ Cotton linters

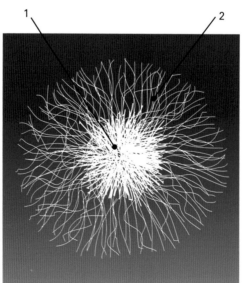

▲ Rag paper is amazingly resistant to traction and pulling apart, since its fibers are long and very pure.

▼ Cotton on the plant

Paper from Rags

In previous times, rag was almost the exclusive source of raw material for paper, along with rope, sandals, and fishing nets. Rag "pickers" and used clothing dealers went from town to town to gather up used textiles. After the rags were taken to the mill, they had to be cleaned. After being removed from the bags or bails, they were winnowed by tossing them into the air to get rid of the initial fine dust. This was one of the hardest tasks in the mill, and it was done by women. Next, the rags were sorted by quality: first, second, third, and extra-fine. Extra-fine rags were composed of white linen, cotton rag, or the best hemp, which was used to make the highest grade of pulp.

Once the rag was sorted, buttons were removed and seams opened up before it was shredded in the chopper. The rags were then put through the fulling mill—a wooden, conical, or hexagonal apparatus that spun around a longitudinal axis, tumbling the rags inside to remove and settle out the impurities.

In the final phase, the rags were submerged and softened for several days or weeks in a fermenting pit holding plain water or water containing calcium carbonate. Then the rags were rinsed and the milling began, which involved all of the steps through refining—such as a series of vats if a hammer process was used, or adjustments to the Hollander beater.

Today, rag paper is considered to be any paper that is made from recycled textiles and whose fibers are of plant origin (cotton, linen, hemp, etc.). Cotton rag, one of the most popular papers, is one of the purest, toughest, and most pleasant-feeling because the fiber is pure cellulose. No chemical processing is required to isolate its fibers, since they are naturally present in the plant as individual units. Of the two types of fibers that cotton contains (long ones and linters), the long ones are preferred by industry for spinning. As a result, recycled textiles are a raw material for high quality cotton papers.

When making paper from rags, fabrics can be used whose composition is known for certain. Contemporary fabrics are frequently a combination of cotton or linen and synthetic fibers, but there are a lot of 100% cotton garments as well that can be recycled to make high quality paper.

For example, later in the book you'll see how to make paper from old blue jeans that are made of pure cotton. Since they are worn almost everywhere in the world, it makes them easy to find. This type of rag can be thrown directly into the beater without any chemical processing. The final product is strong and beautiful.

▲ Museo-Moli Paperer de Capellades (Cataluña, Spain). This shows the area of the mill where the rag was treated before being processed.

▲ Torn and beaten rag

◄ A bale of rags as it would have arrived at the mill. The services of the rag "pickers" and used clothes dealers was indispensable to paper production until wood fibers came into use.

▼ Shredding rags

Comparative Study of Various Papers

In the paper industry, precision instruments are used for measuring the quality of paper. These devices measure all the facets of a paper's strength, its absorption qualities, its opacity, and many other special characteristics that can be created during manufacturing. An artisan who makes paper by hand doesn't own these instruments, so it's necessary to observe the paper by looking at it and touching it. It's a good idea to examine the sheets after experimenting with various fibers and jot down your observations during the process. This makes it possible to work more deliberately without unnecessary experimentation.

The following is a table of observations recorded when experimenting with different fibers. It illustrates some of the many observations that can be made. The page that follows contains three very different paper samples and an analysis of each.

Parameters of Observation

Wet Characteristics *(while the sheet is being formed)*
- Drainage: the time that it takes for the water to drain from the mould
- Set: how hard it is to tear the sheet, or whether it stays intact
- Lay: after pressing, how hard it is to separate the paper from the felt without it tearing

Dry Characteristics *(air drying, with no pressing)*
- Smoothness: stability of the surface, wrinkles, bumps, etc.
- Contraction: papers shrink as they dry, and the degree of shrinkage can be calculated
- Deckle edges: description of the shape of the edges
- Resistance to traction, creasing, tearing, and rupture: depending on our criteria, each paper can be rated from 1 to 10
- Rattle: describes the sound produced when the paper is struck with by hand or shaken, and the relative stiffness of the sheet
- Feel: the sensation (whether smooth, rough, etc.) upon touching the sheet and the body (specific volume of the sheet)

TABLE OF OBSERVATIONS: E (eucalyptus) / K (pine kraft) / C (cotton linters) / H (hemp)									
FIBER		100 % E		100 % K		100 % C		100 % H	50 % C 50 % E
REFINING TIME		20 mins.	1 hr.	15 mins.	45 mins.	15 mins.	30 mins.	30 mins.	30 mins.
WET	Drainage	fast	slow	fast	very slow	very slow	slow	very slow	fairly slow
	Set	delicate	easy	secure	very delicate	delicate	easy	secure	easy and secure
	Lay	delicate	easy	secure	very secure	not very secure	a bit delicate	easy	easy
DRY	Smooth	slight bumps	bumps on edges	many bumps	distorted and twisted	minor waves	minor waves	bumps and waves	very few wrinkles
	Contraction (%)	2	10	5	30	0	0	5	0
	Deckle Edges	few and short	hard	few and short	very hard and serrated	rare	irregular, with threads, and cracked	irregular, with threads	few; subtle, elegant
	Resistance — Trac.	6	8	9	10	6	7	10	7
	Resistance — Rupt.	5	6	9	10	6	7	9	6
	Resistance — Fold	6	7	8	9	5	7	8	7
	Resistance — Tear	6	7	8	9	5	6	8	6
	Rattle	dull sound, like cardboard; stiff	sharp sound, almost metallic	sharp metallic sound; very stiff	very metallic sound; very stiff	sound like rag; not very stiff	sharp sound like rag; not very stiff	very metallic sound; very stiff	sharper sound, like cardboard; not very stiff
	Feel	smooth, soft	satiny, a little hard; rough	soft and hard; not much body	very hard and smooth	soft and rough; lots of body	not very rough; lots of body	smooth and hard	very smooth body; fairly soft

▶ The first of these samples is a paper made with 50% eucalyptus wood fiber and 50% recycled newspaper print that was lightly refined by beating it for 20 minutes in a Hollander beater. It weighs approximately 180 grams (6.3 oz). Synthetic pigments and fragments of recycled pulp were used for color. With respect to the fibers, this paper is low quality, even though it is attractive. It is intended for use as stationery.

▶ The second sample, of better quality, is a sheet of paper made of 100% pine kraft pulp that was thoroughly refined by beating it 45 minutes in the Hollander beater. It is colored with a natural blue pigment. It weighs 130 grams (4.6 oz), and the surface is vellum. Note the rattle characteristics when you manipulate it—it is crisper than the first sample and makes more noise. It is an exceptionally resistant paper.

▶ The third sample is a result of experimenting with a paper composed of 100% cotton linter. The natural white of the cotton has been preserved, and the weight of 300 grams (10 oz) is greater than in the previous samples. It has a soft feel and less rattle than most papers, which makes it well suited for use as a ground for artistic purposes.

O N THE following pages, paper is made step by step in a series of projects that incorporate the theoretical information we've already covered. There are nine very different projects, some of which are more complex and labor-intensive than others. These projects will give you an idea of the limitless creative possibilities for making unique paper by hand, paper that fits your personal tastes.

Paper isn't usually made as an end in itself; rather, it is made with a purpose in mind. For example, it might be used as a ground for painting or other media, or as a medium for creating a sculptural work. As a result, every kind of paper relates to a subsequent application. Even though each exercise focuses on the specific process of making a sheet, the final purpose guides the choice of materials and technique, whether the paper is to be used to create a notebook, album, envelope, card, or artist's print edition. These projects are merely suggestions, and you may find that you will want to add your individual vision to them.

Step by Step

Personalized Paper with Relief and Watermarks

*O*ne of the first things that handmade papermakers often do is create a trademark for their paper—a symbol or sign that identifies it and serves as a personal seal lending individuality to each sheet. Traditionally, this is done by incorporating watermarks into the moulds.

In the following project you'll learn how to make a watermark for stationery and a relief version of it for marking dry envelopes and any other type of ground. The engraver Jordi Catafal collaborated on the creation of the relief. One hundred percent recycled cotton rag paper is used here because its surface takes an imprint well, but you could use any other type of paper made with long fibers.

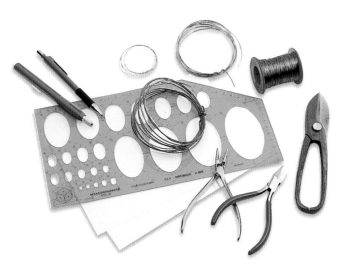

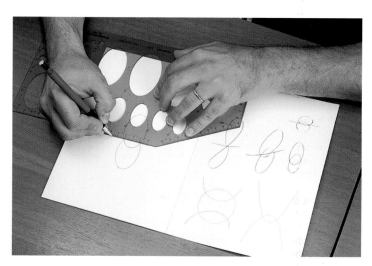

▲ For drawing the watermark, you'll need some drawing utensils including a draftsman's stencil, a drawing pencil, and a few colored pencils. You'll also need aluminum, copper, or brass wire around 2 mm in diameter, some of the finest galvanized copper, steel, brass, or iron thread available, a pair of pliers, and wire cutters.

▲ **1.** To create this design, draw two interconnecting ovals that create a third space between them. (We chose an ellipse because it's an energetic, dynamic, and balanced shape that presents no technical problems for making the watermark. This design is based on the ideas of the contemporary philosopher Eugenio Trias.)

▼ **2.** Once the drawing is completed, trace a pattern for the wire using colored pencils that indicates where each piece of wire will begin and end. (Most marks can't be made using just one section of wire; it usually it takes two or more.)

◄ **3.** Using the pattern as a guide, shape each piece of wire with the pliers, cutting it as needed. (We used aluminum wire because it's easy to work with.)

▼ **4.** Position the wire pieces on the drawing, and adjust the shape as needed to make it as accurate as possible.

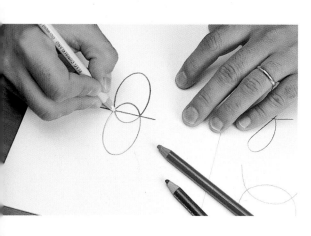

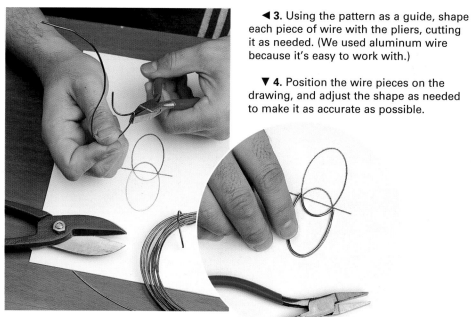

▲ 5. This is the finished watermark. It's very important to avoid creating complicated twists and turns or very tiny closed spaces. Otherwise, the form may pull the pulp away when it's removed in the couching, especially with thinner papers, ruining the paper and image.

► 6. Use stiff metal thread (as fine as possible, but still strong, to keep it from breaking) with no needle to sew the watermark into place. Space the stitches far apart. (It's a good idea to do a preliminary sketch on the mesh before sewing on the watermark. This will help you figure out the best position for the mark, depending on where you want it to appear on the paper, and where to thread it on.)

► 7. Here, the watermark is sewn into place on the mesh. Note the distance between the loops of thread.

► 8. Test the watermark to see how it leaves its impression on the couched sheet. Make sure that the spaces and shape are correct and no pulp is removed from the sheet when the mould is removed.

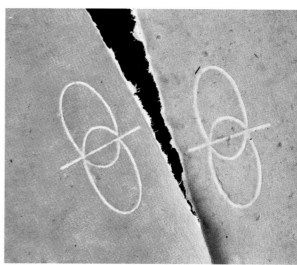

▲ 9. After the paper dries, it shows the beauty of the watermark when held up to light.

▼ 10. You can use a 1 mm copper plate to make a relief of the watermark to emboss dry paper. Begin with the same watermark design, and enclose it in a rectangle of the appropriate size before cutting the plate to fit.

▼ 11. Clamp down the plate securely on the corner of a table with a piece of felt on top to protect it. Use a flat file to bevel one of its faces all the way around so that the edges don't damage the paper when pressure is applied.

▼ 12. Cover the back of the plate with plastic sealing tape. Flip the plate over and coat the beveled face of it with a thick layer of satin-finish black engraving varnish, making sure to apply it to the bevel as well. (Because of the tape and varnish, acid will only be able to reach the plate through the drawing that you'll apply to it.)

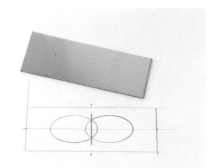

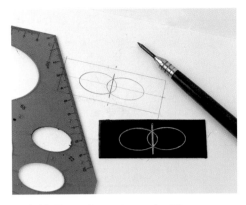

▲ **13.** Draw the watermark with an engraving point to expose the copper plate. This procedure doesn't require much pressure, but it does require accuracy. If you make a mistake, varnish the plate again and start over.

▲ **14.** Make an acid solution (15% hydrochloric acid, 5% potassium chlorate, and 80% water) in a plastic tub. Use it only in a well-ventilated area. When making the solution, always put in the water first, and carefully add the acid to prevent splattering. If you get acid in your eyes, flush them immediately with plenty of running water and then get to a hospital. (If you wish, wear a pair of safety glasses to help protect your eyes from this possibility.) Submerge the plate gently into the acid bath, being careful not to get acid on your hands.

▲ **15.** Leave the plate in the bath for around 24 hours, or until about 80% of the depth of the plate is etched. Remove the plate with stainless steel or plastic tongs so that acid doesn't get on your hands.

◄ **16.** Thoroughly rinse off the plate with plenty of water.

► **17.** Remove the varnish with a solvent. (In this case, it is rubbed gently with sawdust soaked in kerosene.)

◄▼ **18.** Before impressing the design, lightly moisten the paper with a little water from a spray bottle, distributing the water evenly. (If you plan to imprint a long run of papers, moisten them all at once, and keep them wrapped up in a plastic bag so they stay damp.)

▶ **19.** Draw a template of the envelope or other paper shape on a piece of heavy board to use a guide for placing the relief on the paper to be imprinted. Then position and outline the plate. Keep in mind that the dry watermark will emboss the paper in the reverse from how it looks on the plate.

▶ **20.** Place the paper on the template, being careful to keep the plate within the positioning lines.

▶ **21.** Cover the paper with a felt to spread out the pressure and keep from damaging the rollers of your press.

▶ **22.** Next, feed the assembled pieces through the rolling press. (Experiment with this step in advance to adjust the pressure of the cylinder. There shouldn't be so much pressure that the relief damages the paper.)

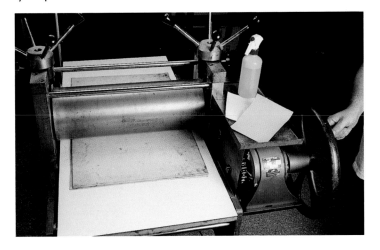

▶ **23.** Once the paper has been fed through, pull back the felts, and check the imprint to see if the pressure of the press needs to be increased. Feed the paper through a second time, making sure that the paper doesn't change position, or you'll create a double image.

▲ **24.** Close-up of the imprint

▶ **25.** Finished paper showing examples of the watermark in the paper and the watermark dry embossed on the paper

Recycled Paperboard

*R*ecycled paper can be very attractive if you use pulp combinations from different sources. In the following project, you'll learn how to recycle the information overload from pages of a telephone book to make interesting paper.

The result is a double-sided heavy paperboard with considerable rigidity. Every sheet is double thickness, and one of the sides is made up of very hard, stiff kraft pulp. The colored paper pulled from the phone books is also fairly strong and stiff. The paperboard produced can be used for such things as file folders, notebook covers, or an album.

▲ **1.** First, you'll prepare the pulp for one of the sides of the sheet by recycling pages of a telephone book, including the thicker, colored ones, such as those containing city maps and ads. (These colored pages combine nicely with the long lists of phone numbers.) To begin, rip the pages out and tear them into pieces of a uniform size.

▲ **2.** Next, put the paper shreds into water and allow them to soak for awhile so that the fibers soften up and are easier to grind up. Then proceed to grinding, and continue this process until the pieces are of a fairly small size.

▲ **3.** To prepare the cardboard pulp, shred it and leave it to soak in water a bit longer because it is thicker and often contains glues. When it has softened up, grind it.

◄ **4.** The ground up cardboard pulp should have a more or less uniform appearance, based on how long it has been in the vat or beater. Here, it has been adequately ground up without refining it, so that it retains some small pieces of cardboard to lend texture to the surface of the paper.

◄ **5.** To create a contrast in color between the two sides of the paperboard, add dark red pigment to the cardboard pulp.

▶ **6.** Stir the pulp until the color is evenly distributed and as dark as you wish. Here, we've created a rust-colored pulp that will set off the yellowish tones of the other pulp.

◀ **7.** Since two sheets form each piece of paper-board, you'll have to work with two vats of pulp. If possible, use a different mould for each vat so you don't have to clean the mould every time you change vats. Use the first mould to couch the recycled phone book pulp on the felt.

▶ **8.** Prepare the second mould with a layer of cardboard pulp. Carefully position the mould, lining it up along the same edge of the felt that you used to couch the first layer.

▲ **9.** After you couch the second layer, you can clearly see the two layers and their relative thickness. Once they're pressed, they'll join together perfectly to form a single sheet.

▶ **10.** The colors of the paper reappear after it dries.

Palm Paper Album

*E*xtracting fibers from the plant world is one of the most interesting facets of papermaking. The results are always good since the paper is strong and has a natural texture that can be seen and felt. In this project, you'll use a chemical processing system to make paper from plants. We chose to use palm, but the steps could be followed to use other plants by varying the cooking times.

   You'll also learn how to bind the paper into a blank book that can be used for any purpose that you like, such as an autograph book, sketch book, or journal. This medieval binding system was originally used with parchment. The paper is stitched directly to the cover. This binding shown was done by Àngels Arroyo.

▲ Palm paper has an attractive color and texture. It's hard, with considerable rattle, but soft to the touch.

▲ **1.** The light color of this "white" palm is due to the fact that the trees grow in special conditions of light and temperature. (You don't have to use white palms, any common palm will do for making paper.)

◄ **2.** To begin, cut the plant into pieces. An electric chipper is quick and convenient, especially if you're making a large quantity of paper. Avoid cutting yourself on the blades and wear goggles to protect yourself from pieces of plant that might fly out.

◄ **3.** If you don't need a chipper, you can use simple pruning sheers to cut up the branches and leaves.

▼ **4.** Next, place the cut plants in a kettle filled with water. There's no precise formula for the amount of water, but the more water there is, the more of the chemical agent (caustic soda) you'll need to add in the next step.

▲ **5.** Add the caustic soda to the pot. (For a 10-quart [9.5 liter] pot like the one pictured, the amount is about 1 teaspoon.)

▶ **6.** Cook the plants, keeping the heat under the pot for several hours.

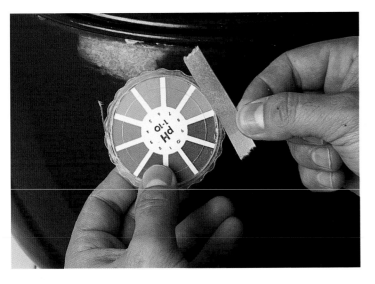

▲ **7.** Caustic soda is an alkaline agent that works quickly in this process if the pH is kept around eight. Check the pH several times during the process. When it goes down to six or seven, add more soda.

▶ **8.** Note: Halfway through the process of cooking these plants, we noticed that the stalks weren't responding to the cooking. After cooking them for three hours, they were still hard. For this reason, the boiling was interrupted, the contents rinsed, and the thick stems taken out and cut apart. Then the process was started again using smaller pieces.

◄ **9.** By touching the fiber, you can tell if it is ready. It shouldn't be resistant to tearing and should be soft, like some edible green that has been boiled. If it's slightly harder than that, it's not a serious problem, since it will be broken down further in the Hollander beater or a blender. When the fiber is cooked enough, take it off the heat, pour it into a colander over a sink (or tile floor with a drain), and rinse it thoroughly with water.

◄ **10.** Rinse the fiber until the escaping water is clear, since the water that you used for boiling the plants now contains lots of impurities, as well as the soda.

► **11.** The next step is the milling, whether in the Hollander beater or a blender. Making this type of pulp is fairly labor intensive and takes longer to break down into fibers. This photo shows the same pulp at four different stages of the process, from left to right: boiled, ground, beaten, and refined. In the center, note the color and the texture of the refined pulp in aqueous suspension.

▼ **12.** Use a mould to form a sheet from the pulp. You'll notice that it drains quickly. The couching and laying are also easy, since the fibers stick together well.

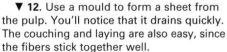

▼ **13.** Dried and pressed finished paper with rustic deckle edges. (Sizing was added during the refining stage.)

► Some of the materials and tools you'll need are: ruler, awl, hammer, metal ruler, scissors, craft knife, pencil, bookbinder's board, wide clear tape, book cloth, bone folder, hemp or linen thread, and leather for the cover. (We chose black suede for our cover.) Binding glue is the best type to use, but any other type of white glue will also work. You'll need a roller/applicator and pan for the glue.

▼ **14.** To begin making your album, fold the palm sheets one by one to form the signatures.

► **15.** Assemble the signatures. (For the sake of lightness, we used six signatures of two sheets each.)

▼ **16.** Cut out two bookbinder's board covers slightly larger than the pages of your album (see page 129 to get an idea of the size of our cover). Cut out a cardboard template the width of the spine. Use a ruler to mark off the spaces between the signatures where you'll insert the needle when you sew them together. From bookbinder's board, cut out a spine that is the same size as the template.

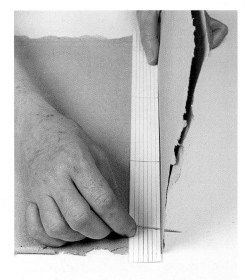

► **17.** Before sewing the signatures, punch holes at the points indicated by the template so that they will line up with the spine.

◄ **18.** Next, you'll make the cover, beginning with the preparation of the leather and book cloth. First, use the roller to apply a smooth, uniform coat of sizing to the book cloth. Work quickly and decisively because binding glue dries very rapidly.

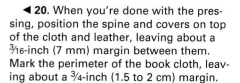

▲ **19.** Place the cloth facedown on the leather. Rub the cloth with a rag to even out the glue and eliminate lumps and bubbles. Press it under weights for an hour to assure stability and successful bonding.

◄ **20.** When you're done with the pressing, position the spine and covers on top of the cloth and leather, leaving about a ³/₁₆-inch (7 mm) margin between them. Mark the perimeter of the book cloth, leaving about a ³/₄-inch (1.5 to 2 cm) margin.

► **21.**Use the craft knife and a metal ruler to cut the book cloth and leather to size.

► **22.** After cutting the cloth, cover the entire surface of it with glue, making sure that there are no lumps or areas without glue.

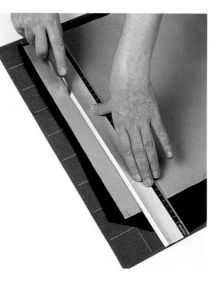

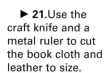

▲ **23.** Place the covers and spine onto the glue-covered surface, using a ruler as a guide to make sure that everything is lined up and spaced properly. (Leave a ³⁄₁₆-inch [7 mm] margin between the spine and covers.)

▲ **24.** The space between the spine and covers allows room for opening and closing the album.

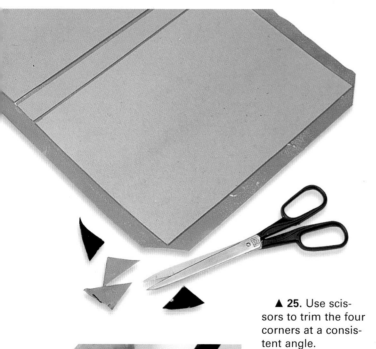

▲ **25.** Use scissors to trim the four corners at a consistent angle.

▲ **26.** Fold the margin of the cover material inside of the covers.

◀ **27.** Use a bone folder to press down carefully on the corners—the areas that are most likely to become distorted.

▶ **28.** Cut out a piece of book cloth and leather slightly smaller than the exact dimensions of your cover, and laminate them together as you did in step 19. Allow them to dry.

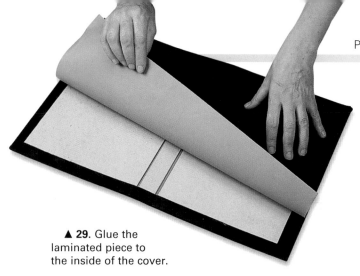

▲ **29.** Glue the laminated piece to the inside of the cover.

▶ **30.** Run the bone folder in the grooves between the spine and cover so that the two faces adhere together well on the inside.

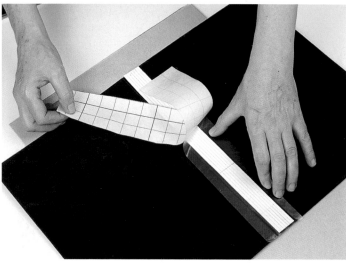

▲ **31.** Use clear tape to position the spine template on the inside of the cover.

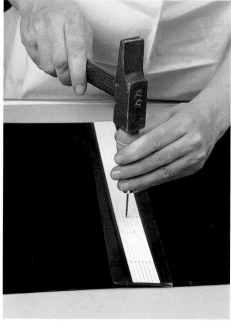

◀ **32.** To facilitate sewing, use the awl to punch the spine along the same points that you marked in step 17, to allow for the needle's passage. Make sure to pierce every hole, opening a passageway between the two faces of the cover. When all the holes are done, remove the template. Now the album is ready to be sewed together.

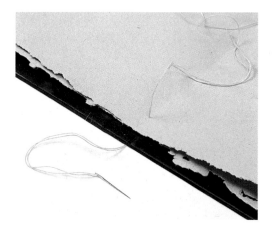

▲ **33.** Next you'll use the needle and thread to sew the signatures to the spine, starting at one end and working from the inside to the outside of the book. To begin, leave a loose tail thread about 4 inches (10 cm) long inside for tying a knot.

▲ **34.** When the thread is on the outside, send it back again through the next hole on the same line.

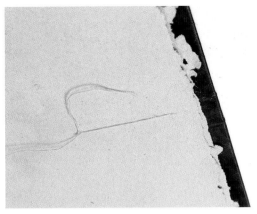

▲ **35.** When the thread is back inside, pull it taut and tie a simple knot with the tail reserved for this purpose. Then push the thread through the next hole up and pass it back into the signature, this time through the last hole.

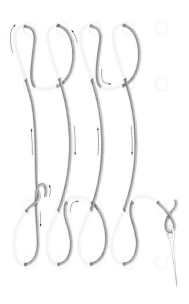

▲ A sketch of the stitching. Note how the thread goes in and out of the book, sometimes through the same hole twice.

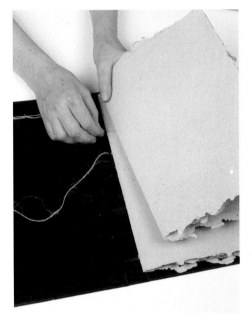

▲ **36.** From the last hole on the inside of the signature, go down to the previous hole and pass the thread to the outside of the spine and change rows. This time the thread will be inside the spine before entering the next signature.

◄ ► **37.** The finished album. The stitching pattern is clearly visible on the spine (left). A front view displays the attractive combination of colors and materials (right).

Stationery Made from Paper with Colored Threads

*U*sing color in handmade paper is one of the best ways to give the finished product a unique appearance. Paper made using colored thread is one of the most traditional and attractive kinds. The threads appear as very fine lines in the paper.

In this project, we'll incorporates threads in the pulp that is then made into stationery. Adding threads results in decorative papers with random designs, attractive colors, and a pleasant feel. The pulp we used is a mixture of fibers: 25% eucalyptus, 25% pine kraft, and 50% cotton linters, with an internal sizing added in the beater.

◄ For textile inclusions, select some fibers of plant origin, such as pita, jute, hemp, cotton, or linen—but never wool. The colors of these fibers contrast nicely with the pulp used to make the sheet.

► **1.** Begin to prepare the fiber by cutting the yarn into segments.

◄ **2.** Since the yarn is often too thick for the paper, use your fingers to fray and separate the cut-up segments.

◄ **3.** If you're using spools of thread on cardboard tubes, you can cut the tubes up the middle to create threads of the same length.

▲ **4.** Once the spools have been cut apart, separate the threads.

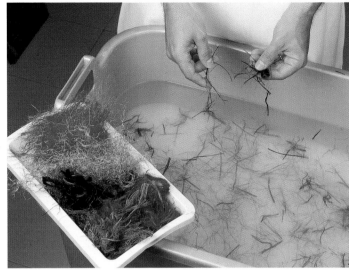

▲ 5. The textile fibers you've selected are your color palette.

▲ 6. Mix the threads into the pulp. You'll notice that they blend in right away.

◄ 7. The threads integrate uniformly with the pulp when the sheet is being formed.

▲ 8. Change the color of the pulp by adding an initial color. (You'll change the color of the pulp again later by adding a second color.)

► 9. Place an envelope template on the mould, followed by the deckle.

▲ **10.** Shake the mould and deckle back and forth to help integrate the threads. (The deckle retains the pulp and water, facilitating the process.)

▲ **11.** Remove the deckle carefully, trying to avoid allowing drops of pulp to fall on the paper.

◄ **12.** Tip the mould to allow any surface water to drain off, then lift off the template. Couch and press the envelope on a felt. (Make as many envelopes as you wish from the yellow pulp.)

► **13.** Add red color to the pulp, until you arrive at a deep-colored orange. Make as many envelopes with this pulp as you wish, and couch them.

▼ **14.** Note: If you wish, you can use a thick template for your envelopes made from a piece of plastic (about ¼-inch [6 mm] thick) instead of using a deckle and template.

▼ **15.** After you've pulled the sheet, carefully remove the template from the envelope.

▼ **16.** If you want to create a deliberate design, add the fibers on top of the paper while it's in the mould or on the felt. You can also add a bit of diluted pulp on top to better incorporate the threads.

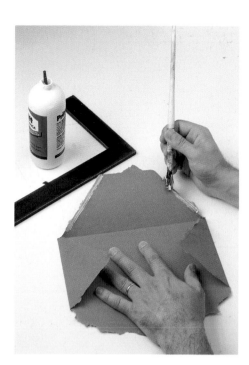

▲ **17.** After the envelope has dried, use a T-square to guide you as you fold it, making sure that all the corners are perfectly square.

▶ **18.** Coat the contact edges of the envelope with glue, press them together, and allow them to dry.

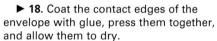

▶ **19.** Make envelope glue by dissolving one part gum arabic, one part starch, and four parts sugar in water (in that order). Boil the mixture for a couple of minutes.

▶ **20.** Coat the edges of the flaps with this glue. Allow the glue on the flaps to dry flat for about an hour without touching one another.

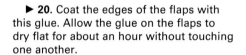

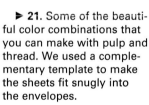

▶ **21.** Some of the beautiful color combinations that you can make with pulp and thread. We used a complementary template to make the sheets fit snugly into the envelopes.

Writing Paper and Envelopes Made from Jeans

For many centuries, Western paper has been made using pulp from recycled rags. Rag paper is very strong and pure, especially if made from cotton. This project will show you how to make rag paper using 100% cotton blue jeans. Because of the long cotton fibers, the resulting paper is strong and has a lot of body.

The envelopes you'll make play off of the idea of denim jeans, including added labels, decorations, and machine-stitched seams. Teresa Collado contributed to this project.

▲ **1.** Before cutting up the blue jeans, cut off the rivets, buttons, labels, and zippers. Save the labels for later.

▲ **2.** Rip the fabric into strips slightly less than 1 inch (2.5 cm) wide.

▲ **3.** Next, use scissors to cut out squares of a consistent size.

▶ **4.** Since there are many types of blue jeans in many color shades, it's helpful to sort them by color for making pulps of the same color or creating specific mixes.

◄ **5.** Before grinding up the rags, weigh the dry material in order to calculate the right amount for the batch. (The dry material will equal the weight of the paper. There is no loss of material.)

▲ **6.** Use plenty of water to grind up the rags in the Hollander beater. If needed, boil the rags beforehand using caustic soda to help break them down. If you're preparing pulp with a blender, this is an important step to take.

◄ **7.** The milling begins with very little pressure. Add some bleach to lighten the color as well as weaken and extract the fibers.

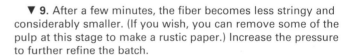

▼ **9.** After a few minutes, the fiber becomes less stringy and considerably smaller. (If you wish, you can remove some of the pulp at this stage to make a rustic paper.) Increase the pressure to further refine the batch.

▼ **8.** After the fiber has been in the beater for a few minutes, it's stringy, and the threads are visible.

▲ **10.** Pour the pulp into the vat. At this point, it's already refined, free of threads, and uniform in color—a pale blue that's very typical of this type of fabric. (If you put lots of bleach into the pulp and allow it to sit for a few days, it will turn pure white, but the fiber will weaken.)

▲ **11.** If you didn't add sizing during the refining process, do it now.

▶ **12.** Stir the pulp well to eliminate lumps.

◀ **13.** In sheets formed from this type of pulp, the drainage is good and the result is very stable.

▼ **14.** Two papers made from blue-jean pulp at a couple of different stages in the refining process. The more rustic of the two has some unrefined fragments and color variations.

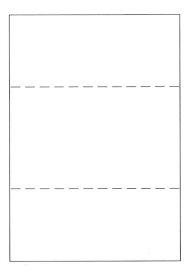

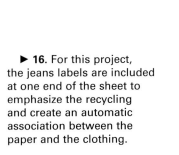

◄ 15. The first combination of envelope and sheet will be a simple fold with decorative zigzag lines imitating a seam (as seen in the illustration).

► 16. For this project, the jeans labels are included at one end of the sheet to emphasize the recycling and create an automatic association between the paper and the clothing.

▲ 17. After placing the label, cover it with a little pulp to make it bond to the paper.

▲ 18. Use your finger to carefully remove the excess pulp covering the outside edge of the label before couching and drying the sheet.

◄ 19. Some examples of dried papers with labels attached.

▲ **20.** Use the sewing machine and different colored thread to experiment on separate sheets of the paper with various stitches and tensions. Determine the tension that won't tear the paper.

◄ **21.** To make the envelope, fold the sheet so that the label sticks out.

▲ **23.** On another piece of paper, sew a zigzag stitch around the edge to create writing paper.

▲ **22.** Sew the sides of the envelope together and around the top edges with a zigzag stitch.

► **24.** The final result. The labels lend a distinctive note to the project.

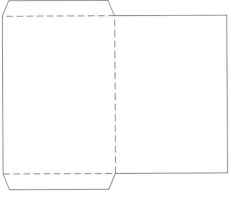

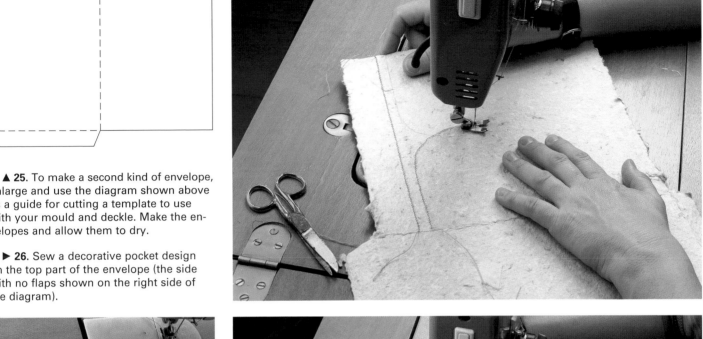

▲ 25. To make a second kind of envelope, enlarge and use the diagram shown above as a guide for cutting a template to use with your mould and deckle. Make the envelopes and allow them to dry.

▶ 26. Sew a decorative pocket design on the top part of the envelope (the side with no flaps shown on the right side of the diagram).

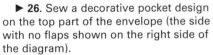

▲ 27. After you finish, fold the flap over so that the sides of the envelope meet, and trim the loose threads. Fold the bottom part of the flap up and over the envelope.

▲ 28. Starting at the bottom of the envelope, sew all the way around the edge and top and come back down the other side to join the open seam.

▶ 29. The finished envelope resembles a blue-jeans pocket.

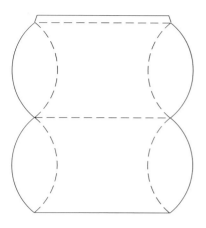

▲ **30.** Enlarge this template on a piece of paper to use as a guide for cutting your envelope from handmade paper.

► **31.** For this envelope, select some sheets of paper to use. (We used less-refined paper, so it would have more texture.)

▲ **32.** Cut the paper out according to the diagram. The sketch must be precise, so that everything matches up perfectly. Use a bone folder to fold the paper along the curved dotted lines seen in the diagram.

▲ **33.** Next, apply bookbinder's glue to the flap. Fold the sheet in half along the scored line, and fold the glued flap to the back of the piece. Fold in the curved flaps.

▼ **34.** The finished envelope/ packet makes an ideal box for enclosing a small gift.

◄ **35.** To make a third envelope variation, enlarge the diagram at the left.

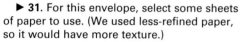

► **36.** Use the template as a guide to cut out the circle from a paper of your choice, and use a bone folder to fold the paper along the dotted lines.

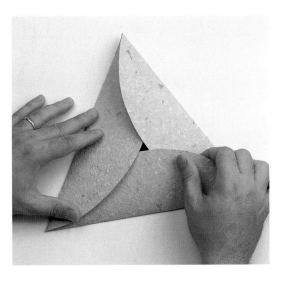

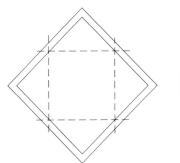

◄ 37. Fold in the flaps, as shown.

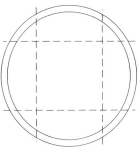

▲ 38. These diagrams show more alternatives for making envelopes.

▼ 39. The papers that you use to make interesting envelopes can be made from recycled blue-jean pulp combined with other recycled pulps, such as cardboard and newsprint.

► 40. Experiment with different combinations of papers and folds.

► 41. Examples of finished envelopes that provide unique ways to send letters, cards, and invitations, or store special papers

Postcards, Cards, Bookmarks, and Decorative Shapes

*W*hen you intend to make a paper of unusual format or size, it's common to draw and cut out that format or shape on another, larger sheet of paper. Or, as you discovered in the previous project, it's possible to make special papers right in the mould by using separating templates as well as watermarks. The benefit is that you don't lose the natural deckle edges of the sheets.

In this project, you'll learn to make decorative paper shapes using moulds. You'll also learn to make paper that can be used as postcards, calling cards, business cards, or personal note cards. Papers shaped in the mould can be used as bookmarks, ornaments, and so forth. All the papers shown are made in the vat from 80% eucalyptus fiber and 20% kraft, with added internal sizing. They are colored using natural pigments.

We've also incorporated flowers and algae into some of the papers because these materials add to the romantic flair of the projects.

◄ This exercise was done in the Ca l'Oliver paper mill of Sant Quintí de Mediona in Barcelona, Spain. The flowers and plants used in the project were gathered from the environment right around the mill.

► 1. Moulds for cards and postcards. A watermark made using a coarse grating threaded onto the mesh creates divisions for making several small sheets in a single couching.

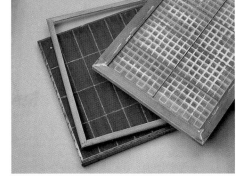

◄ 2. Natural pigments create pastel colors.

▼ 3. With a prepared form, 36 cards can be made in a single couching.

◄ 4. After couching the paper onto a felt, you can see clearly the divisions on the moist paper. This size card is ideal for notes, invitations, or postcards.

◄ Algae makes a nice addition to the pulp.

► Dried flowers collected from the grounds of a garden also make good inclusions.

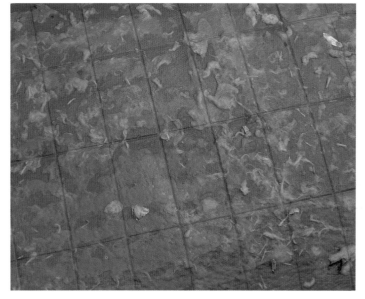

▲ **5.** If incorporating dried flowers, soak them in the vat for a couple of hours so that they blend into the paper more easily.

▶ **6.** To make varied cards from the same sheet, you can couch another shade of pulp with inclusions on top of a monochrome background. To create a gradation of color on each side of the mould, pick the pulp up by partially submerging one edge of the mould and then the other.

▲ **7.** Once the paper is couched on the felt, you can see the colorful effect. Keep in mind that the colors won't be as bright when the paper dries.

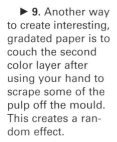

◀ **8.** You can create double-sided postcards by adding another color of pulp on top of the first. To do this, line up the second couching precisely.

▶ **9.** Another way to create interesting, gradated paper is to couch the second color layer after using your hand to scrape some of the pulp off the mould. This creates a random effect.

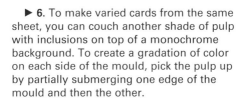

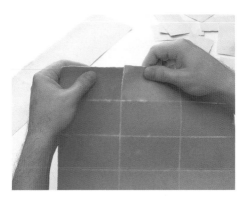

▲ **10.** When the cards and postcards finish drying, tear them along the lines to create frayed edges.

▶ **11.** Examples of cards and postcards that have been dried and pressed

◄ For making papers with different shapes, deckles with cutouts are used. These are made from ¼-inch (6 mm) tempered hardboard.

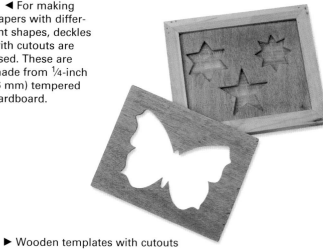

► Wooden templates with cutouts

◄ **1.** Remove the deckle and template to reveal the shaped paper.

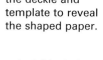

► **2.** Maximize the use of space on the felt by placing shaped pieces close to one another.

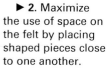

▼ **3.** Ornaments can be made with a shaped deckle. Remove the deckle carefully to avoid damaging the areas with tight angles.

▼ **4.** You can create hangers for ornaments by placing wires between layers of pulp.

◄ **5.** Flower petals and wild grasses make nice inclusions in decorative papers.

► **6.** To make bookmarks with decorative flower petals, arrange them by hand on the pulp while it's in the deckle. Then, put a bit of diluted pulp on top.

► **7.** You can also add inclusions directly to the vat so that they are distributed randomly.

► **8.** Pulp with grasses and petals added in the vat works nicely in the shape of a butterfly.

▼ **9.** Finished butterfly and bookmarks

► **10.** Stars can be placed on a table as decorations, used as coasters, or made into ornaments. The shapes that you can create with cut deckles are unlimited. You can make shapes to suit any holiday or celebration.

Large Format Paper

*I*n this project, we'll demonstrate how to make large-format white paper used in the fine arts as a ground for drawing, painting, and engraving. Paper of these dimensions requires larger equipment and the hands of two people.

Fine arts paper must be made from very high quality fibers, preferably of non-wood origin, with a neutral pH. For the size shown in this project, the paper should be heavy weight (around the thickness of paperboard). Because it's unlikely that you'd find this size equipment on the market, you'll probably need to construct a vat and mould to meet your size requirements if you want to undertake making large-scale paper. The mould must be ribbed to support such large paper. Lluis Morera and Oriol Mir collaborated on this project.

▶ **1.** Before submerging the mould, stir the contents of the vat thoroughly to get rid of lumps and even out the fiber.

▲ **2.** Submerging a mould of this size is fairly simple when done by two people.

◀ **3.** Gently lower the mould to the bottom and then lift it straight up, keeping it even.

▶ **4.** When a mould of this size is lifted from the vat, it's very heavy because of the quantity of water and pulp it holds. (Oriental paper makers use a system of ropes and harnesses on bamboo anchored in the ceiling to help reduce the resistance of the mould. The bamboo is strong, but flexible enough to help in raising or lowering the mould.)

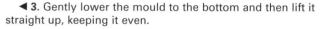

▲ **5.** Shake the pulp back and forth in the mould to evenly distribute the fibers. (To keep the pulp from running between the mould and the deckle, hold the two together with a clip at the midpoint along the edge.)

▼ **8.** The frame has to be lifted up with the same motion used in couching: from one side, as if opening a door.

▲ **6.** Remove the clips before carefully lifting off the deckle, avoiding drips that could happen easily because of its large size. Center the mould and support it well before couching the sheet.

◀ **7.** Laying paper of this size must be done decisively but smoothly, and it's easy to create wrinkles. After the mould is positioned, press down firmly on the ribbed frame, which is essential in this type of mould.

▶ **9.** This type of paper is usually dried on a covered horizontal rack in the open air. (Hanging it vertically would probably tear it because of its weight.) If you have a large press, the paper can be pressed; otherwise it can be allowed to dry on the felt. Since the weather was sunny and calm when we made our paper, we decided to leave it outdoors to soak up the sun's warmth and light for a couple hours, in Oriental fashion.

Accordion Book with Colored Images and Text

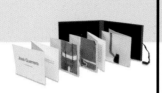

*F*rom an artistic point of view, working with colored pulps can be very interesting. In this project, you'll see how to make a series of pictorial compositions based on an existing design. The final result is never the same as the starting point, since there is always the characteristic lack of clarity in images made with paper pulp. This factor is also part of the charm.

Many graphic artists hire papermakers to create limited editions of their works that can take the form of freestanding pieces or art books. The following project is based on a commission from an individual client—the result is a portfolio holding interpretations of pictures and text by José Guerrero.

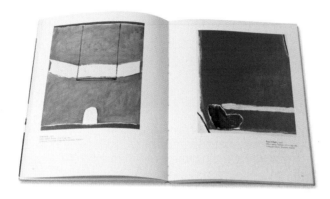

▲ **1.** The content for the portfolio was selected from the artist's catalog. Since the size of the images fit the size of the portfolio, no enlargements were needed.

▶ **2.** To reproduce a pictorial work, you'll need sheets of clear acetate and a permanent marker to make templates. The template shown indicates where the red will go.

▼ **3.** Use a craft knife is used to cut out the template, leaving the areas to be colored exposed.

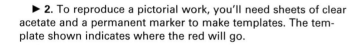
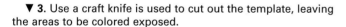

▼ **4.** Place the clear mask on the mould and hold it in place with the deckle.

► **5.** Instead of submerging the mould in the vat, pour the pulp on the mould without moving the acetate. (The color of the pulp was arrived at by experimentation. We used cotton linter fiber that was refined for 25 minutes so it would pass easily through the nozzles of the applicator bottles used later.)

► **7.** Couch the colored pulp onto a very clean felt. Keep in mind that the final image will be reversed right and left.

▲ **6.** Remove the mask to reveal the image.

▼ **8.** Push down on the mould to solidify the layer.

▼ **9.** Use an applicator bottle full of pulp to touch up any errors in color or detail.

◄ **10.** Before applying each color, check the shade by squeezing it on a sheet of paper. The pulp should leave no halo of pigment around it. If there is too much pigment, it will run and spoil the composition. The photo shows the adjustments made to the blue pulp.

◄ **11.** Use other colors to fill in the open spaces of the composition as needed, based on the original design. (In this case, we're filling in the blue heart.)

▼ **12.** For our particular design, we cut out another acetate template to define the boundaries of the design. We positioned it on the felt around the pulp. Then we couched a sheet of white pulp on top of the template to fill in the white areas. When the mould and template are removed, the boundaries of the design are revealed and the excess white pulp is lifted off with the acetate.

► Another way to work is to use applicator bottles to directly apply pulp to the felt (instead of with the mould). This method allows you see the image in the right perspective and avoid the couching; however, the colors will run together somewhat because the water and pulp build up with no way to drain.

▼ **13.** To give the paper more volume, you can couch another layer of pulp onto the back of the design. First, place the mould to determine where to couch this sheet, and mark its position with two pieces of tape for registration purposes.

▼ **14.** For extra strength and support, laminate the final sheet on the back. The drawing remains centered.

▲ **15.** Three designs in colored pulp based on Guerrero's original ones. These interpretations retain the artist's sense of color and space. The edges are trimmed so that the pieces can be put together with the text in an accordion-style book.

▶ **16.** A long sheet of heavy paper (paperboard) will be used to support the images and text in an accordion shape. The text is therefore trimmed to the same height as the pictures.

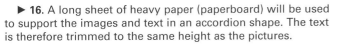

▲ **17.** The text is attached to the paperboard with spray adhesive, which prevents wrinkles and distortion. As each block of text is added, the previous one is masked with a piece of paper to prevent smudging.

▲ **18.** Position the text after the adhesive has been applied.

▲ **19.** The pictures, which are on heavier paper than the text, are attached with white glue.

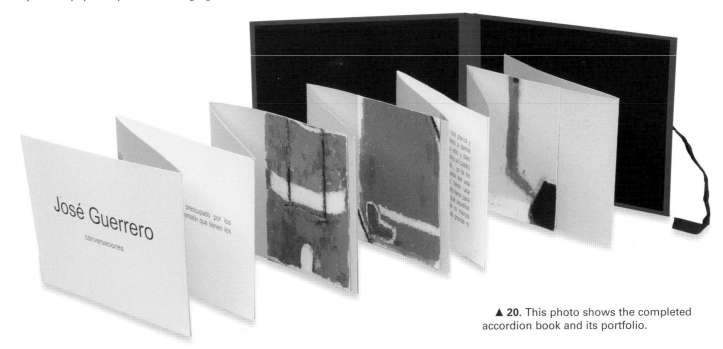

▲ **20.** This photo shows the completed accordion book and its portfolio.

Rustic Notebook with Textured Paper

*H*andmade paper references the world of nature. It's very intriguing to think of creating a booklet or book from plants, without buying any kind of material. You merely need to boil the plants, extract their fiber, and use it to make paper. The book's binding can be very simple. In this project, you'll see this sort of notebook produced. Texture dominates the look of the piece—burlap is used to lend texture to the inside pages, which are made from abaca and hemp fibers. The covers, which have the same fibrous composition, contain rice straw, kozo fiber in its natural state, and stems from a bush. The originality of this notebook lies in the fact that its covers are constructed entirely of materials integrated into a single sheet of paper.

▼ The following materials are needed in addition to the paper pulp: rod-shaped stems from a bush, strips of *kozo* plant fiber, and rice straw.

◄ **1.** The first step is to make a very thick sheet that will serve as the outside of the cover. This particular pulp is colored with a dark pigment to contrast with the pages placed inside. Rice straw is mixed into the dark pulp in the vat.

◄ **2.** While the sheet is forming, you can add even more straw to create texture.

▼ **3.** Once the cover is couched on the felt, center rod-shaped stems along the spine, one for each signature of paper you plan to put in the notebook.

▼ **4.** We've placed four rods along the spine for four signatures of paper.

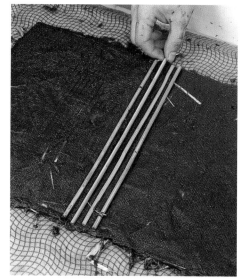

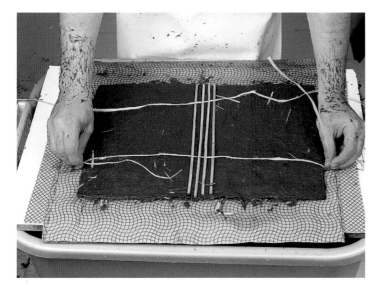

▲ **5.** Next, place strips of plant fiber across the cover to serve as ties for the notebook later. *Kozo* plant fibers are long and very strong, but you can also use string.

▲ **6.** Couch a smaller sheet of the pulp over the first to serve as reinforcement and integrate the ties and rod into the cover.

▶ **7.** Then couch another contrasting color of pulp onto the cover that is the same size. This sheet is the inner lining, like the fly-leaf of a book.

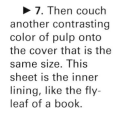

▲ **8.** The cover as it looks after pressing and drying under weights to keep it from distorting. This process also helps to adhere the various layers.

▶ **9.** For the inside pages, we made white paper textured with burlap. The pulp is made from abaca and hemp, whose long, strong fibers withstand the process of making texture. Pages made with this pulp have a certain amount of rattle, and they wrinkle quite a bit as they dry.

▶ **10.** Allow the sheets to dry without pressing to encourage wrinkling and retain a more rustic, textured appearance.

▲ **11.** Fold the signatures. We used four signatures with three sheets each for a total of 24 pages.

▲ **12.** Stack the signatures. Use a ruler to guide you as you make notches with a craft knife at even intervals along the folded edges. These cuts mark spots where you'll pierce the needle and make it easier to pass through.

▲ **13.** Make corresponding marks for placement of the needle on the inside of the cover.

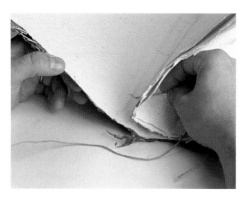

▲ **14.** Using hemp thread and a large needle, begin the stitching. Push the needle outward from inside the first signature in preparation for bringing the thread around the first rod in the spine.

▲ **15.** Push the needle back up through the inside cover, around the rod.

▲ **16.** Push the needle through the same hole that it entered in the signature, and knot the end. (Knotting is not repeated until the very end of the stitching, in the last signature.)

▼ **17.** Pass the thread to the outside through the next hole so that it encircles the rod on the outside of the spine.

▼ **18.** Bring the thread back through to the inside, then pass it under the previous stitch to reinforce it.

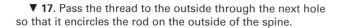

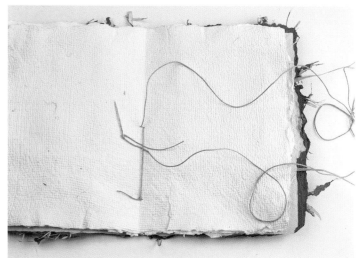

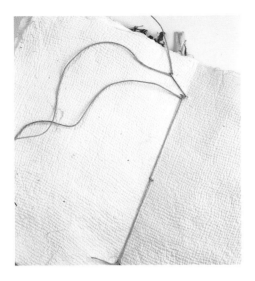

▲ **19.** When the whole signature has been stitched in place, push the needle back to the outside of the signature.

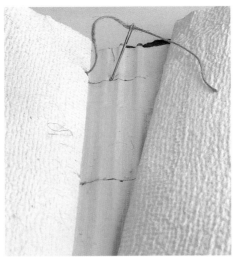

▲ **20.** Before stitching in the next signature, push the thread to the outside of the cover and around the next rod.

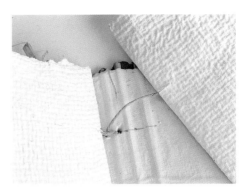

▲ **21.** The rod is encircled in preparation for sewing in the next signature. Repeat the steps of sewing in the signature, this time working from top to bottom.

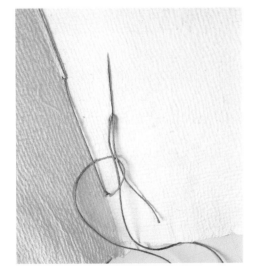

◄ **22.** In the final signature, the last pass of the thread has to be knotted to strengthen the stitching.

▲ **23.** This is what the spine looks like after the booklet is sewn together, and you can see the thread that goes around the rods, although it blends in well because it is the same color as the rice straw. (Another option is to use a contrasting color, since the stitches can also be an attractive feature.)

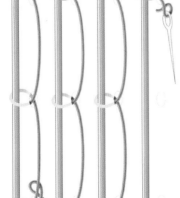

◄ Follow this diagram for stitching.

► **24.** The finished notebook with ties that give it rustic character

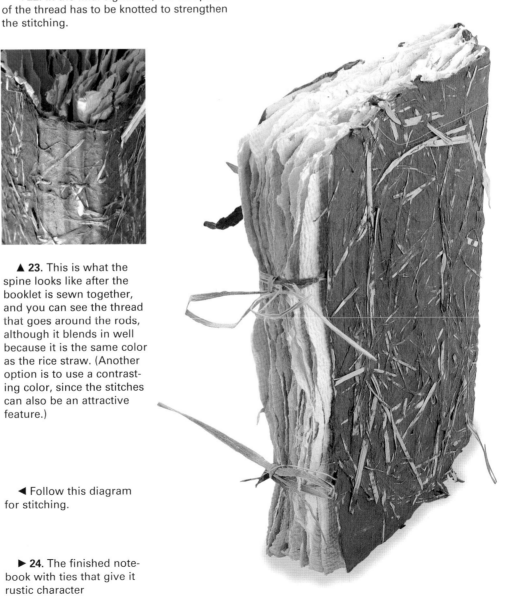

a

Alum: Potassium sulfate and aluminum used to thin animal sizing and as a mordant.

b

Bale: A bundle containing eight to ten reams of paper.

Base plate: Toothed iron or steel part located on the base of the Hollander beater. The beater roll, which is equipped with grooves or blades, turns on the base plate.

Batch: Quantity of pulp or rag put in the cylinder or the Hollander beater.

Beating: The step taken in papermaking prior to refining and after shredding.

Beater: A generic name for equipment in which rag is ground up or pulp is refined. Historically, beaters were made of stone.

Beater roll: Wood, iron, or stone cylinder of the Hollander beater. The beater roll is equipped with bronze or steel blades arranged parallel to the rotational axis. It is used to extract the fiber from rags and for making pulp.

Beating time: The time required for extracting the fibers from rag and refining the pulp in the Hollander beater.

Bible paper: Lightweight, fine, opaque paper commonly made from linen, suitable for printing on both sides.

Bleaching: Eliminating the natural color of the fiber and whitening it, using chlorinated water and calcium chloride in the pulp. Formerly, this was done using calcium carbonate or bleaches made from ashes prior to developing the present chemical methods.

Body: A set of 25 sheets of paper or five signatures. This term is commonly used with reference to the specific volume of a sheet of paper, since paper has more or less strength based on its volume.

Broken sheet: Paper that's rejected because of some defect: spots, burrs, tears, creases, fingerprints, smudges, rope marks, irregularities, dimples, thick spots, etc.

Burnisher: A device used to burnish and impart a shine to the paper. In olden times, this was done by hand using a highly polished stone. Later a pantograph was used, which had an agate or an onyx attached to one end.

c

Calender: Each of the metal cylinders over which the paper passes to acquire a satin finish.

Caustic soda: Sodium hydroxide used for whitening rags and extracting fiber

from plants. Sodium carbonate is also used when a less aggressive method is desired.

Coucher: The worker who takes the mould from the vatman and transfers the sheet of paper to a felt on the couching table before returning the mould and placing a new felt on top of the sheet.

Couching: The action of moving the sheet of paper from the mould to the felt.

Couching bench: Wooden board on which the couching process is carried out. The old style couching benches had two reinforcements underneath that acted as a guide when sliding it toward the press.

Couching station: The wooden framework that serves as a protective apron next to a vat where the vatman stands. The coucher also stands in a vertical niche between the vat and the press.

Crease: An occasional wrinkle in a sheet that's impossible to eliminate, caused by defects in the felts, or because the lay man didn't do a good job of placing it onto the laying table.

Cross: T-shaped piece of wood used to hang up paper on the lines in a drying loft.

Cutter: Blade set up vertically and secured to a table that is used for ripping up rags, removing buttons, and undoing seams.

d

Deckle: A wooden frame that's placed onto the mould for making paper that allows water retention and controls drainage while the sheet is forming.

Deckle edge: Unevenness on the edges of handmade sheets of paper.

Deckle paper: Paper with uneven edges called deckle edges. The best quality paper of this sort is made from linen using the best rags.

Digester: Wooden apparatus in the shape of a cone or hexagon that turns around along an axis and agitates the rags placed inside it to remove impurities.

Drying rack: Wooden rack with legs used to hold the newly made paper to be hung up to dry.

f

Felt: Rectangle of seamless wool fabric or other material used as a base for laying paper.

Fermenting pit: A space in the paper mill where rags were placed to ferment before grinding them up.

Filler: Products added to the paper to fill pores, improve quality and rattle, and give it such characteristics as heavier weight, improved ability to take print, opacity, and shine (kaolin, sizing, starch, talc, resins, etc.).

Format: The dimensions of sheets of paper determined by the size of the book, magazine, or other printed matter. Every format has a name, and in olden times a watermark was used for identification.

h

Hog: Wooden paddle traditionally used for mixing the pulp in the vat or beater.

Hollander beater: (also cylinder mill, fiber extractor, refining cylinder). Machine for extracting fibers invented in Holland around 1670. It consists of an oval-shaped reservoir with a divider in the middle, a beater roll, a base plate, and a hood.

k

Kraft paper: Very tough paper made with sulfated cellulose. It's used for bags, wrapping paper, etc.

l

Laid paper: Sheets of paper made one by one with classical laid moulds, or on a machine with a drum engraved with a mesh pattern.

Lay: To separate the paper from the felt.

Laying bench: Traditionally, a wooden device about 30 inches (80 cm) high onto which the lay man places the paper sheets as he separates them from the felts. It's shaped like an easel or an inclined plane with two legs in front and a movable arm in the rear that holds it up.

Lay man: Vat worker whose job it is to separate the sheets from the felts when they come out of the press and place them on the couching table.

Linters: Short fibers that cling to the cotton seed after ginning.

Loft: (also drying room) A location that is almost always on the top floor of a paper mill where the paper is hung up to dry.

m

Mesh: The weave of the mould.

Mill: To grind up rags for making paper pulp.

Mould: The device used for forming sheets of handmade paper. It consists of a wooden frame with mesh or a strainer that spreads out the pulp at the same time that it facilitates water drainage. Normally the mould is reinforced from underneath with a series of strips.

p

Papermaker's tears: The most common defect in handmade paper produced when the deckle is taken off the mould and drops of water fall on the paper, or when a very wet mould is passed over the paper on the felts.

Picardo: Named after its inventor, this round papermaking machine produces paper very similar to deckle-edged vellum.

Pilcher: A large folded felt that is placed on top of a stack of papers to be pressed.

Press: The essential tool in a paper mill that's used to press the paper, extract excess water, and make the sheets smooth once they're dry.

Pulp: Mass of broken-down fibers or rags for making paper, whether refined or not. Once it is refined it is called refined pulp, and it can be of various consistencies, depending on the size of the fibers.

q

Quire: One-twentieth of a ream; traditionally 24 or 25 sheets of paper of the same size.

r

Rattle: A special sound made by the finished sheet of paper when it's shaken. Each fiber has a distinct rattle. Papers with external sizing can be identified by their hard, metallic rattle. A related characteristic is the stiffness of the sheet.

Ream: Traditionally, a set of 500 sheets of paper, or 20 quires.

Refine: To reduce the paper pulp to a finer consistency.

s

Signature: A folded sheet of paper used as a unit of a book.

Sizing: One of several products added to paper to make it water-resistant. Animal glue or gelatin is extracted from the hides, viscera, and bones of animals. Plant glue can be made from flour or rice starch and from resin. There are also synthetic glues that work well as sizing.

Stamper: Block of wood or iron set up vertically and activated in the same way as beating hammers were for the purpose of beating pulp. It moves on a metal yoke or plate.

t

Trim: To even up the paper using scissors, a knife, or shears.

Trimming bench: In a traditional mill, a box on four legs on which a pack containing one or two reams of paper was secured using ropes and a screw press. A knife was used to trim the edges of the paper.

v

Vat: Wooden, stone, plastic, or other receptacle in which paper pulp is made.

Vat man: A worker in a traditional paper mill who uses the mould to make a sheet of paper from pulp in the vat.

Vellum: Smooth paper/parchment, traditionally made from animal skin, made using a fine metal fabric or machine-made mesh that leaves no marks on the paper.

w

Watermark: The papermaker's sign made from very fine silver or copper wires threaded onto the mesh or mesh of the mould. In factory-made papers the watermark is stitched onto the mesh of the drum.

Web: Continuous machine-made paper that comes in large rolls.

Winnowing the rags: Removing rags from bags or bales and tossing them into the air to get rid of dist. This was one of the hardest jobs in a tradtional paper mill, and it was done by women.

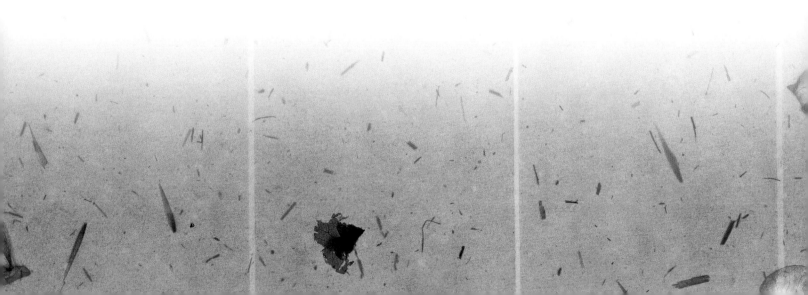

Bibliography
and Acknowledgments

AA.VV. *"El moli paperer de Capellades."* Quadern de didactica i difusio (vol. 5), Museum of Science and Technology of Cataluña, Terrassa, 1990.

AA.VV. *El Museu-Moli Paperer de Capellades.* Museu-Moli Paperer de Capellades, Capellades, 1991.

AA.VV. *Innovations and Explorations in Handmade Paper: Twenty Years of Collaboration at Dieu Donné Papermill.* Display Catalog, Dieu Donné Papermill and Gallery, New York City, New York, 1996.

AA.VV. *Papiers et moulins, des origines à nos jours.* Editions Technorama, Argenton-sur-Creuse, 1989.

Bo Rudin, *Making Paper, A Look into the History of an Ancient Craft.* Rudins, Sweden, 1990.

García Hortal, José Antonio. *Constituyentes fibrosos de pastas y papeles.* Department of Textile and Paper Engineering of the Terrassa School for Industrial Engineers, Terrassa.

Heller, Jules. *Paper-making.* Watson-Guptill Publications, New York, 1978.

Hiebert, Helen. *Papermaking with Plants.* Storey Books, Vermont, 1998.

López Anaya, Fernando. *El Papel hecho a mano. Elementos de grabado artístico y composición tipográfica.* Buenos Aires, 1981.

M. de la Lande. *Arte de hacer papel.* Clan, Madrid, 1995.

Shannon, Faith. *Paper Pleasures.* Mitchell Beasley International Ltd., London, 1987.

Turner, Sylvie. *Which Paper?* Design Press, New York, 1992.

Valls, Oriol. *Vocabulari paperer.* Museu-Moli Paperer de Capellades. Capellades, 1999.

Thanks *to Jordi Catafal for inspiring me to enter the world of papermaking as a profession some years ago, and for his collaboration as an engraver for this book. To Lluis Morera for his initial lessons and to the entire team at Ca l'Oliver for their support, especially Toni Capellades. To Gail Deery and Loraine Walsh for teaching me techniques I didn't know.*

Thanks *to José Antonio Garcia Hortal and the entire team of teachers and students at the Universitat Politècnica de Cataluña, ETSEI de Terrassa, who for years have devoted lots of energy to researching paper fibers. Without their research and publications, the chapter in this book devoted to fibers would not have been possible.*

Thanks *to Victoria Raval and the team at the Museu-Moli Paperer de Capellades for their incredible work in keeping this craft alive and disseminating it, and for allowing access to their setup for this this book project.*

Thanks *to the Escola d'Arts i Oficis de la Diputación de Barcelona, their team of professors, especially those who work with me in the publishing arts department (J. Catafal, M. Monedero, J. Cambras, and M. Vallmitjana).*

Thanks *to my students, because I have learned a lot from them. I especially want to thank the students from the 2000–2001 course for their analysis of fibers reflected in this book. Thanks to the Council of Barcelona, especially to the Education Division, for allowing certain projects to be carried out in the school.*

Thanks *to Ramon Serera for his help in the historical section, to Teresa Collado and Angels Arroyo for the contributions in all the projects, and for their good work and creativity.*

Thanks *to Joan Soto for his kindness—a source of ideas beyond the technical—and for his wisdom about photography resulting from many years of experience.*

Thanks *to the entire team of professionals at Parramon Publishing, especially to the editorial director, María Fernanda Canal, for the confidence, enthusiasm, and support she gave me.*

Thanks *to my wife Gemma and my children, Gerard and Georgina, for their tremendous patience and support.*

Thanks *to my sister Mercè for having introduced me to the fascinating world of publishing and to Montserrat Guasch for her generous collaboration throughout the entire time that I worked on the book.*

Thanks *to my parents for supporting my decisions and my love of art.*

Josep Asunción